STATE OF THE UNIVERSE 2008

DATE DUE	RETURNED

Martin Ratcliffe

STATE OF THE UNIVERSE 2008

NEW IMAGES, DISCOVERIES, AND EVENTS

Springer

Published in association with
Praxis Publishing
Chichester, UK

Martin Ratcliffe FRAS
Wichita
Kansas
USA

Front cover illustration: This infrared image from NASA's Spitzer Space Telescope shows the Helix nebula, the remains of a star that once looked like our Sun. When sun-like stars die, they puff out their outer gaseous layers. These layers are heated by the hot core of the dead star, called a white dwarf, and shine with infrared and visible colors. In Spitzer's view of the Helix nebula, infrared light from the outer gaseous layers is represented in blues and greens. The white dwarf is visible as a tiny white dot in the center of the picture. The red color in the middle of the eye denotes the final layers of gas blown out when the star died. Image courtesy NASA/JPL-Caltech/K. Su (Univ. of Arizona).

Back cover illustration: **(Top)** This image taken with the Advanced Camera for Surveys on NASA's Hubble Space Telescope depicts bright, blue, newly formed stars that are blowing a cavity in the center of a star-forming region in the Small Magellanic Cloud. Image courtesy NASA, ESA, and the Hubble Heritage Team (STScI/AURA) - ESA/Hubble Collaboration. **(Middle)** This composite image shows the Crab Nebula, the remains of a colossal stellar explosion called a supernova. The Chandra X-ray image is shown in light blue, the Hubble Space Telescope optical images are in green and dark blue, and the Spitzer Space Telescope's infrared image is in red. Image courtesy NASA, ESA, CXC, JPL-Caltech, J. Hester and A. Loll (Arizona State Univ.), R. Gehrz (Univ. Minn.), and STScI. **(Bottom)** This image by the Chandra X-ray Observatory shows Cassiopeia A, the youngest supernova remnant in the Milky Way. The red and green regions show material from the destroyed star that has been heated to millions of degrees by the explosion. Image courtesy NASA/CXC/MIT/UMass Amherst/M.D.Stage et al.

SPRINGER-PRAXIS BOOKS IN POPULAR ASTRONOMY
SUBJECT ADVISORY EDITOR: John Mason B.Sc., M.Sc., Ph.D.

ISBN: 978-0-387-71674-9 Springer Berlin Heidelberg New York

Springer is a part of Springer Science + Business Media *(springeronline.com)*

Library of Congress Control Number: 2007936093

Cover design and cartoons: Jim Wilkie
Typesetting and design: BookEns Ltd, Royston, Herts., UK

Printed in Germany on acid-free paper

CONTENTS

■ *Pink Pleiades. The Seven Sisters, also known as the Pleiades star cluster, seem to float on a bed of feathers in this infrared image from NASA's Spitzer Space Telescope. Clouds of dust sweep around the stars, swaddling them in a cushiony veil. The 19th-century poet Alfred Lord Tennyson described them as, "glittering like a swarm of fireflies tangled in a silver braid." The spider-web-like network of filaments, colored red in this view, is made up of dust associated with the cloud through which the cluster is traveling. One of the parent stars, Electra, can be seen extreme right, with the other three brightest members of the cluster above right and below left of center. Additional stars in the cluster are sprinkled throughout the picture in blue. Image courtesy NASA/JPL-Caltech/J. Stauffer (SSC/Caltech).*

Preface

STATE OF THE UNIVERSE 2008

WELCOME TO the second volume of a popular level annual review of astronomical discoveries, State of the Universe 2008. In these pages you'll find easy-to-read, bite-sized sections of news from beyond our solar system, covering gamma-ray bursts and black holes to the newest extrasolar planet research and more, over a full 12-month period (April 2006 to March 2007).

In addition to the selected review of the top stories, I've invited ten authors and astronomers to contribute broader overviews of specific topics that are at the forefront of current research, placing the chronological news stories into a broader context.

I am particularly thankful to the research astronomers who gave of their valuable time to write these reviews, in between writing grant proposals, presenting papers at conferences, and actually observing with the world's best telescopes. Without their passion for communicating what they do to a broader audience, this book would be far less interesting.

I am pleased to welcome back for a second year, Jim Kaler, whose review of top news from the 12-months provides a versatile and insightful segue from the news bites to the rest of the invited articles.

Seasoned science writer, Carolyn Collins Petersen, joins us for the first time. She reviews the latest news in low frequency arrays, and the prospects for major discoveries from studying long radio waves.

Planet formation and star formation are intimately linked, and I've invited four experts to review the current state of our knowledge in this rapidly developing field of research. Stephen Strom and Luisa Rebull review star formation, with the latest that the Spitzer Space Telescope is telling us. James Graham and Paul Kalas focus on their newest discoveries inside planetary debris disks.

The latest discoveries from the Chandra X-ray Observatory, one of NASA's Great Observatories, are reviewed by Chandra expert, Wallace Tucker. Observing at the other end of the electromagnetic spectrum is one of America's national treasures, the National Radio Astronomy Observatories. NRAO's Public Information Officer, David Finley, reviews recent discoveries made from its outstanding facilities.

The Sloan Digital Sky Survey (SDSS) has already revolutionized astronomy. One of the SDSS astronomers, Timothy Beers, contributes a feature reviewing another exciting year of discovery.

Hubble Servicing mission Project Scientist, Chris Blades, from the Space Telescope Science Institute, writes a timely and authoritative review of the exciting mission to update the Hubble Space Telescope, expected to occur in late 2008.

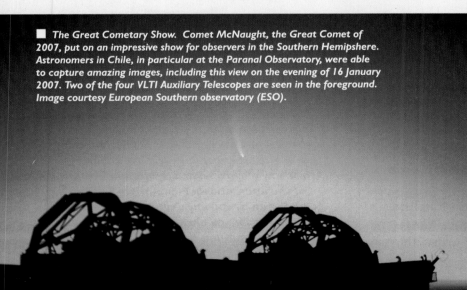

■ *The Great Cometary Show. Comet McNaught, the Great Comet of 2007, put on an impressive show for observers in the Southern Hemipshere. Astronomers in Chile, in particular at the Paranal Observatory, were able to capture amazing images, including this view on the evening of 16 January 2007. Two of the four VLTI Auxiliary Telescopes are seen in the foreground. Image courtesy European Southern observatory (ESO).*

■ *The starburst galaxy NGC 1313. The very active state of this galaxy is evident from this composite image, showing many star formation regions. A great number of supershell nebulae – cocoons of gas inflated and etched by successive bursts of star formation – are visible. Images obtained with the FORS1 instrument on one of the 8.2-m Unit Telescopes of the ESO Very Large Telescope. The data were extracted from the ESO Science Archive and fully processed by Henri Boffin (ESO).*

Jerry Nelson, whose bright revolutionary idea led directly to the construction of the twin 10-meter Keck telescopes, writes about the 400th anniversary of the telescope. He offers a unique insight into the exciting future of giant telescope construction and adaptive optics.

Alexei Filippenko, an award-winning teacher at UC Berkeley and a leading researcher of one of the teams that discovered the accelerating universe, introduces us to that greatest of mysteries, Dark Energy, in a highly readable article.

The book begins with my annual review of selected top news stories, complete with web links for further reading. I encourage you to take the internet journey; it will enrich your reading, and even link you to the original research papers. This book is a gateway to enormous resources, and is enough to keep you busy for an entire year, that is, until the third volume.

No book is possible without family support, and this one is no exception. For many weekends over six months, "the book" has been a constant presence at home, and for that I am eternally grateful to my wife, Shawn, for her love and active "you need to write" encouragement.

Martin Ratcliffe

Wichita, Kansas
August 2007

Martin Ratcliffe is Director of Professional Development for Sky-Skan, Inc., a digital planetarium manufacturer. For 12 years he has written the monthly night sky column for Astronomy magazine, and is a former President of the International Planetarium Society. Following an astronomy degree from University College London, he has spent 20 years running various planetariums and teaching astronomy.

■ One of four 1.8-meter telescopes, known as the Very Large Telescope Interferometer (VLTI) Auxiliary Telescopes (or ATs, for short), seen in silhouette against a twilight sky viewed from the summit of Cerro Paranal in Chile. The ATs are 100 percent dedicated to VLTI, while the VLT's four main 8.2-meter Unit Telescopes are only intermittently available for interferometric observations. The ATs can be placed on any of 30 possible stations and therefore provide many interferometric baselines. Image courtesy ESO.

1 A year in NEWS...

The Human Astronomers were keen to catch up on the news: but a little nervous about waking Bunny!!

Each month a flood of new results pours in from the world's observatories. Here *Martin Ratcliffe* reviews the major highlights, selected from the hundreds of news reports released between April 2006 and March 2007. Each story profiles the main news and provides weblinks, allowing exploration of further details, images and the actual research articles, bridging the gap between short news bites and the original science.

...and PICTURES
APRIL 2006 - MARCH 2007

APRIL 2006

4 April 2006

Galaxies Align Forming String of Beads

Astronomers have found that spiral galaxies are not randomly oriented in space. They appear to be preferentially aligned. The dramatic results come from a careful study of two huge galaxy surveys, the Sloan Digital Sky Survey and the Two-Degree Field Galaxy Redshift Survey, combined with precise locations of known voids, large spaces lacking in bright galaxies.

Throughout the universe, galaxies are known to be distributed along filaments surrounding large, bubble-like voids, creating a web-like large scale structure. Predictions made by supercomputer modeling of the universe show that following the Big Bang, this labyrinth-like structure forms as a natural consequence of the gravitational interaction of dark matter. Galaxies are the visible "tip-of-the-iceberg" to this primordial dark matter distribution.

Fred Hoyle, in 1951, and Jim Peebles, in 1969, studied the theoretical implications of the primordial matter distribution and its effects on the spin of galaxies or, more accurately, their angular momentum. In the prevailing tidal torque theory (TTT), galaxies acquire angular momentum from tidal sheer induced by the distribution of primordial matter. Such modeling predicts galaxies should be oriented perpendicular to the direction of the linear filaments of the large scale structure. Several

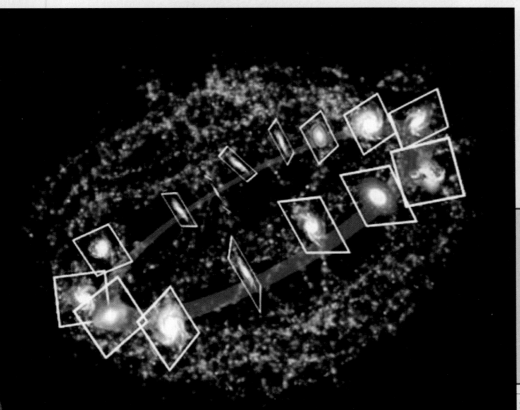

■ **Left:** *4 April 2006. The spin axes of spiral galaxies appear to be aligned with the large scale filaments that surround giant voids. Theory predicted such alignments due to the way infalling primordial gas pooled to form the first galaxies, but observational proof was absent until now. Image courtesy Gabriel Pérez Díaz, MultiMedia Service (IAC).*

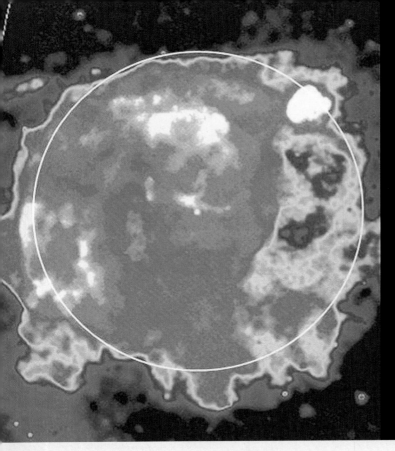
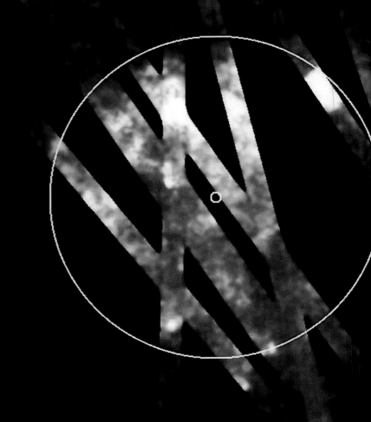

groups had searched for the alignments in the 1980s and 1990s. The difficulties lay in the accurate determination of galaxy orientation to the filamentary large scale structure.

The new results represent confirmation of the predictions. A team led by Dr. Ignacio Trujillo of the University of Nottingham, including astronomers from the Instituto de Astrofisica de Canarias (Spain), developed a successful analysis.

"We found that there is an excess of disk galaxies that are highly inclined relative to the plane defined by the large-scale structure surrounding them," said Dr. Trujillo. "Their rotation axes are mainly oriented in the direction of the filaments." The discovery opens the way to a deeper understanding of the formation of galaxies from the large scale structure of the universe.

http://www.iac.es/gabinete/noticias/2006/
m03d31e.htm
http://www.sdss.org/
http://xxx.lanl.gov/abs/astro-ph/0511680

■ **Above:** 4 April 2006. A comparison of the Vela supernova remnant from ROSAT (left) and XMM Slew Survey (right), reveals some similar features and some that have apparently changed and are presumably variable. A 3.5-degree radius circle on both images acts as a useful guide to compare features. The small circle on the XMM-Newton Slew image indicates the position of the Vela pulsar. Image courtesy Andy Read (University of Leicester, UK), ESA, the Max-Planck-Institut für Extraterrestrische Physik and the ROSAT Mission.

4 April 2006

XMM Newton Slew Survey Reveals Secrets of X-ray Sky

Space-based telescopes are under great demand to observe as many objects in the sky as possible. However, there is limited time for observing each target, and a significant amount of time is spent slewing the telescope from one target to another. This maneuvering results in lost time for observing. Now, astronomers using the European Space Agency's XMM-Newton X-ray observatory have a new approach to analyze X-rays collected during these slewing periods.

The new technique has resulted in the detection of over 4,000 X-ray bright objects, ranging from close binary stars where matter from one star collides and explodes on the surface of its neighbor, to active galaxies, and outwards to distant quasars 10 billion light-years from Earth.

The XMM-Newton Slew Survey is the deepest hard-band all-sky survey ever performed, probing ten times deeper than all previous surveys. The energy range spans the 0.2 – 2 kev band, matching the ROSAT survey from the 1990s, but also extending to harder X-rays in the 2 – 12 kev band.

Indicative of the kinds of new results being obtained is a slew across the Vela supernova remnant, recording in just a few seconds similar features previously requiring dedicated pointing by ROSAT. Many variable sources are evident between the two images. Another result is a dramatically brighter distant elliptical galaxy, NGC 3599. Located 40 million light-years away, this normally quiet galaxy shows up 200 times brighter in X-rays than in previous observations made by ROSAT. An active supermassive black hole at the galaxy's core may be the culprit.

As UK astronomer Dr. Andrew Read, of Leicester University in England, explains, "Over a quarter of the entire sky has already been covered in the 400 or so slews so far performed ... giving us large scale views of the biggest X-ray objects in the sky". Dr. Read expects the entire sky will be covered during the lifetime of the XMM-Newton mission.

http://www.star.le.ac.uk/~amr30/Slew/
http://sci.esa.int/xmm/
http://arxiv.org/abs/astro-ph/0610070
http://arxiv.org/abs/astro-ph/0512157

5 April 2006

Gemini/HST Survey Reveals Building-block Process in Evolution of Massive Galaxy Clusters

In March 2005 the first results from a survey of near and far galaxies, using the Hubble Space Telescope and the twin 8-meter Gemini telescopes, suggested that current theories of how galaxies evolved from their early beginnings to seniority may require some revision. The problem lies in the fact that if the stars in early galaxies simply got older, they would not look like the galaxies we see in large clusters today. Those first results suggested such "passive evolution" is the incorrect model for the history of star formation within galaxies. New results from galaxy cluster Abell 1689, studied by astronomers with the Gemini/HST Galaxy Cluster Project, support the requirement for new thinking.

The passive evolution model showed that galaxies in the center of large clusters produced stars early in their lifetimes and then aged without further changes. The new results reveal that stars in young distant galaxy clusters are very different from those in older, nearby clusters.

When we see galaxies at different distances we are looking back in time. By comparing the chemical makeup of stars in both distant and nearby galaxy clusters, the chemical evolution of galaxies can be determined. The Gemini

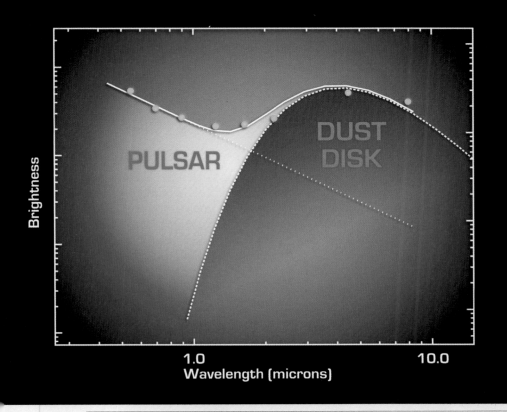

■ **Left:** *5 April 2006. This chart shows the spectrum of the pulsar 4U 0142+61 taken by the Spitzer Space Telescope (SST). It reveals the tell-tale infrared hump that identifies a dusty disk around the dead core of a star, the first ever detection of fallback material following a supernova explosion. The two green dots on the right represent data from SST, and are combined with other data from ground based telescopes. The predicted contributions of light from the pulsar and the dust disk are shown by dotted lines. Image courtesy NASA/JPL-Caltech.*

Multi-Object Spectrograph enables the required analysis of starlight.

"We found the earliest galaxy clusters have a huge variation in the abundances of elements such as oxygen and magnesium, whereas the chemistry of galaxies in the sample of closer clusters appears to be much more homogeneous," explained Dr. Jordi Barr of Oxford University, England.

The observations suggest that the young clusters acquired more elements from mergers with other galaxies. The results also show that lower mass galaxies sustain star formation for about four times longer than their more massive cousins. If a lower mass galaxy which had greater chemical mixing caused by prolonged star formation later merged with a massive galaxy, triggering a new burst of star formation, the chemical makeup would become more homogeneous in the resulting merged galaxy.

Barr adds, "This difference in chemistry proves that the clusters must actively change over time. If the galaxies in the old clusters have acquired a complete "set" of elements, it's most likely that they have formed from the mergers of several young galaxies".

http://hubblesite.org/gallery/album/entire_collection/pr2003001a/

http://arxiv.org/abs/astro-ph/0608150

http://arxiv.org/abs/astro-ph/0601403

5 April 2006

Spitzer Space Telescope Finds Hints of Planet Birth Around Dead Star

In another first for the Spitzer Space Telescope, infrared observations have uncovered a planet-forming debris disk around a pulsar, the compact remains of a supernova explosion. The idea that material can remain around a pulsar following the explosive death of a star is not new. Astronomers call the debris that has insufficient velocity to escape "fallback". Theoretical models suggest that fallback can trigger the collapse of a neutron star to form a black hole, but until now, observational evidence of fallback material has never existed. If the fallback has sufficient angular momentum it may form a rotating disk of material around the pulsar.

Spitzer found evidence of the disk, resembling those found around young stars, from its infrared glow, and weighs in at an estimated ten Earth masses. The estimated lifetime of the disk exceeds the pulsar spin-down period of at least a million years, suggesting the possibility of some of the material sticking together to form planets. The pulsar, called 4U 0142+61, lies in Cassiopeia 13,000 light-years away.

Due to the extreme environment, with harsh X-rays illuminating the disk, any future planets would be unlikely to produce any life. Yet thanks to "fallback", the debris disk does form. "We're amazed that the planet-formation process seems to be so universal," said Dr. Deepto Chakrabarty of the Massachusetts Institute of Technology in Cambridge. "Pulsars emit a tremendous amount of high energy radiation, yet within this harsh environment we have a disk that looks a lot like those around young stars where planets are formed."

These observations provide a missing link between the 1992 discovery of three planets orbiting the pulsar PSR B1257+12 by Aleksander Wolszczan. They were the first planets found outside our solar system.

The team, including Chakrabarty and MIT colleagues, Zhongxiang Wang and David Kaplan, also surveyed four other pulsars and didn't find any disks. However, calculations show that deeper infrared observations with Spitzer could detect one around Puppis A.

http://www.spitzer.caltech.edu/Media/releases/ssc2006-10/index.shtml

http://xxx.lanl.gov/abs/astro-ph/0604076

http://xxx.lanl.gov/abs/astro-ph/0606686

5 April 2006

Swift Observes an Unusual Bang in the Far Universe

Gamma-ray bursts continue to be a hot topic as more and more GRBs are detected by the highly successful orbiting Swift observatory. While many bursts fit a growing pattern, either a massive star collapsing to form a black hole, or a merger of a neutron star and a black hole, once in a while a burst occurs that doesn't fit any of the current models.

Newly announced studies of a gamma-ray burst that occurred on 1 August 2005 provided the first evidence of a different type of burst, perhaps the result of the formation of a rare kind of neutron star. The new burst occurred 9 billion light-years from Earth.

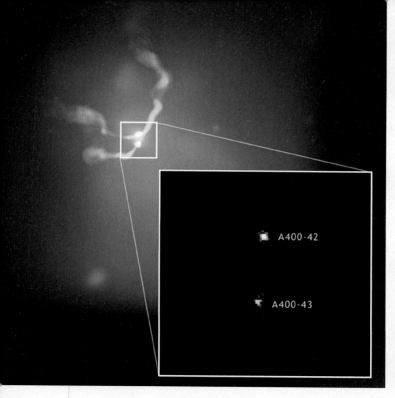

6 April 2006

Study Finds Two Supermassive Black Holes Spiraling Toward Collision

Astronomers using the Chandra X-ray Observatory have detected a pair of supermassive black holes at the center of the galaxy cluster, Abell 400. Eventually, if the two supermassive black holes are orbiting each other, they will spiral in toward a merger.

The two black holes lie at the center of two active galactic nuclei (AGN) that currently lie about 25,000 light-years apart. The AGNs are observed as the double radio source in 3C 75, discovered in the mid-1980s by the Very Large Array in Soccoro, New Mexico. Data taken from the VLA show two pairs of intertwined jets of plasma emanating from each AGN. The oppositely-directed jets are distorted by the passage of the pair of galaxies through the intragalactic hot X-ray gas in which they are embedded. Supermassive black holes are presumed to be the source of jets in AGNs. Careful analysis of the way the two pairs of jets are intertwined, by astronomers from the University of Virginia, Bonn University, and the U.S. Naval Research Laboratory, show that the pair of black holes are gravitationally bound.

Craig Sarazin from the University of Virginia explains, "The jets are similar to the contrails produced by planes as they fly through the air on Earth. From the contrails, we can determine where the planes have been, and in which direction they are going. What we see is that the jets are bent together and intertwined, which indicates that the pair of supermassive black holes is bound and moving together."

In the distant future when the two black holes do collide, gravitational waves will be emitted. Understanding the frequency of such mergers will aid the development of future gravitational wave detectors. In 2002, the Chandra Observatory observed a similar pair of supermassive black holes, but at a much later stage of evolution, in the core of the galaxy NGC 6240. (See also 18 April and 1 May stories.)

http://chandra.harvard.edu/photo/2006/a400/
http://xxx.lanl.gov/abs/astro-ph/0603272

Typically, a burst is followed by a slowly fading afterglow in X-rays and optical wavelengths. Most long bursts are thought to be caused by a black hole swallowing a large star.

The 1 August event (called GRB 050801) did show the usual afterglow but the initial bright flare was missing. A steady emission for over 4 minutes in X-rays and optical eventually led to a normal type of fading. The level of bright emission has not been seen before, and provides hints to the nature of the central engine powering the burst.

"This feature might be explained if we assume that, rather than a black hole, the core of the star has shrunk its mass and its magnetic field into an object known as a magnetar," said Massimiliano De Pasquale of the Mullard Space Science Laboratory (MSSL), University College, London. Magnetars are unusually rare neutron stars with a magnetic field thousands of times stronger than typical pulsars.

The observations were aided by the earliest detection from the ground of the optical afterglow following the burst detection by the orbiting Swift observatory. The automated ROTSE-IIIC telescope at the H.E.S.S. site in Namibia, was imaging the location 21.8 seconds after Swift's initial detection. ROTSE III is a worldwide network of four robotic telescopes built for fast automated response to triggers from orbiting gamma-ray observatories such as Swift.

http://swift.gsfc.nasa.gov/
http://xxx.lanl.gov/abs/astro-ph/0601350

■ *Above left: 6 April 2006. This image is a composite of X-ray (blue) and radio (pink) images of 3C 75 located in the galaxy cluster Abell 400. It shows the twin pair of radio jets immersed in a vast cloud of multi-million degree X-ray emitting gas that fills the cluster. The jets emanate from the vicinity of two supermassive black holes (also shown inset). These black holes are in the dumbbell galaxy NGC 1128 at the center of the giant radio source 3C 75. Images courtesy NASA/CXC/AlfA/D. Hudson & T. Reiprich et al. – X-ray; NRAO/VLA/NRL – Radio.*

■ *7 April 2006. The FORS1 instrument on the 8.2-meter Very Large Telescope acquired this three-color image of the Tarantula Nebula in the Large Magellanic Cloud. The original images were taken through three different narrow-band filters (centered on 485 nm, 503 nm, and 657 nm), for a total exposure time of slightly more than 3 minutes. In the center of the image is the star cluster R 136, whose ultraviolet light from its hot young stars causes the entire nebula to glow. An older cluster of stars, Hodge 301, is located to the upper right of the image. Image courtesy ESO, with thanks to Henri Boffin.*

7 April 2006

VLT FORS Image of the Inner Parts of the Tarantula Nebula

The Tarantula nebula, located at the northern end of the Large Magellanic Cloud, is a spectacular star forming region located 170,000 light-years away. It's one of the largest such clouds visible in detail from Earth, and consequently the target of many telescopes. In a newly-released image by the European Southern Observatory's Very Large Telescope (VLT), one of the largest and most advanced telescopes on Earth, the fine detail of the glowing tendrils of this magnificent nebula are revealed.

The VLT is a set of four 8.2-meter telescopes atop Cerro Paranal in Chile. This image was acquired in 2002 and 2003 using one of the four giant telescopes, and photographed through three selected color filters. The three filters allowed light centered on three wavelengths to be recorded, at 485 nanometers, 503 nm, and 657 nm. With a total exposure of about 3 minutes,

the three individual images were later combined to produce a color picture.

A giant young cluster of hot blue stars called R 136 causes the nebula to glow. Their ultraviolet radiation penetrates clouds of hydrogen gas causing it to re-emit light with a distinctive pink color at a wavelength of 656.3 nanometers. R 136 contains many massive stars, some over 50 times the mass of our Sun. The cluster is estimated to be two to three million years old.

To the upper right of the image is another cluster, called Hodge 301. This is about ten times older than R 136, as indicated by the number of stars that have progressed to the red giant stage.

The images were taken with the FORS1 instrument (focal reducer and low resolution spectrograph), which contains a 2048 x 2046 pixel camera capable of producing a resolution of 0.2 arcseconds on an 8.2 meter telescope.

http://www.eso.org/outreach/press-rel/pr-2006/pr-13-06.html

http://xxx.lanl.gov/abs/astro-ph/0501568

http://xxx.lanl.gov/abs/astro-ph/9910426

18 April 2006

Breakthrough in Black Hole Simulation

Colliding black holes provide scientists with the most extreme environments in which to test Einstein's General Theory of Relativity. Predictions of the appearance of gravitational waves from such encounters have proved extremely hard to calculate, until now. Scientists at NASA's Goddard Space Flight Center (GSFC) in Greenbelt, Maryland, successfully ran the largest 3D astrophysical simulation ever performed on NASA's supercomputer, called Columbia. Previous attempts had caused computer crashes, so the recent success was particularly welcoming to the group who devised new ways of translating Einstein's equations into a code the computer could handle.

The simulations recreate the conditions around a pair of orbiting equal-mass, non-spinning black holes. Each black hole distorts the very fabric of spacetime, and the orbiting pair spiral inwards while generating gravitational waves for many years before the actual merger takes place. Given the number of black hole pairs being discovered (e.g. see 6 April story), being able to simulate them allows greater insight into being able to detect the waves. The Laser Interferometer Gravitational-Wave Observatory (LIGO) is now operational and is at the forefront of the observational challenge of detecting gravitational waves (see article 'The Search for Gravitational Waves … Listening to Space with LIGO' by Laura Cadonati in State of the Universe 2007). Characterizing gravitational waves in the final phase of merging is crucial to extracting a real signal from observational data.

"These mergers are by far the most powerful events occurring in the universe, with each one generating more energy than all of the stars in the universe combined. Now we have realistic simulations to guide gravitational wave detectors coming online," said Joan Centrella, head of the Gravitational Astrophysics Laboratory at GSFC.

In a major step, the repeated simulations show "profound agreement" to within 1 percent. Such repeatability gives Dr. John Baker, the lead author of the research paper, and colleagues, significant confidence in the results. A number of other research groups are following similar approaches using different techniques. It's only a matter of time before the historic first detection of gravitational waves from a black hole merger occurs.

(Note: A second paper by the same authors has been published in early 2007 and the link is provided below. In it they predict current detectors may soon be successful given recent theoretical modeling advances. (See also 6 April and 1 May stories.))

http://www.nasa.gov/vision/universe/starsgalaxies/gwave.html

http://arxiv.org/abs/gr-qc/0602026

http://arxiv.org/abs/gr-qc/0701016

■ **Left:** *18 April 2006. This visualization shows the product of the largest three-dimensional simulations of merging supermassive black holes ever performed using the Columbia supercomputer at the NASA Ames Research Center. Using Einstein's theory of general relativity to predict the nature of gravitational waves from such mergers, the simulation provides a new foundation for the detection of gravitational waves.*
Image courtesy Henze, NASA.

20 April 2006

ESA's ISO Provides the First Clues of Monstrous Stars Being Born

Observations from the now defunct Infrared Space Observatory (ISO) appear to have achieved the first look at a birthplace of massive stars. Two very dense molecular cloud cores each have enough material to produce at least one O-type star along with a cluster of smaller stars.

The regions were found during the serendipity survey performed by ISO during slews between targets, and therefore took advantage of otherwise unused observing time on the orbiting observatory.

One of the cores contains the mass equivalent of 75 suns, and shows signs of infalling gas. The core currently has a chilly temperature of about 16 Kelvin (-256.5 deg Celsius). The second cloud has a slightly lower temperature, a mass of 280 suns, and little turbulent motion, indicating the cloud is near the initial conditions just prior to collapsing to form stars.

The regions where high-mass stars are formed are hidden from view because they lie in dense cores such as these, and require infrared observations to penetrate them. Astronomers have long held that stars are formed within such dark clouds. One particular struggle has been to explain how high-mass stars form? These results suggest what some astronomers suspected: that high-mass stars form in regions of very low temperature and low turbulence.

"This opens up a new era for the observations of the early details of high-mass star formation," says Oliver Krause, Max-Planck Institute for Astronomy, Heidelberg, Germany and Steward Observatory, Arizona. He adds that another major question is "why do some clouds produce high- and low-mass stars, whilst others form only low-mass stars?"

The answers may come with larger and better telescopes. This region and others like it will be high on a target list for Herschel, a 3.5-meter infrared space telescope due for launch in 2008, and for SOFIA, the Stratospheric Observatory for Infrared Astronomy, housed inside a high-flying jumbo jet.

http://www.esa.int/esaSC/SEM8MZNFGLE_index_0.html

http://xxx.lanl.gov/abs/astro-ph/0509710

■ *24 April 2006. The Hubble Space Telescope took this image of the well-known galaxy, M82. Dozens of bright clumps represent starburst regions where intensive star formation is taking place. The violet stellar winds blast filaments of hydrogen gas high above the plane of the galaxy. Image courtesy NASA, ESA and the Hubble Heritage Team STScI/AURA, with thanks to J. Gallagher (University of Wisconsin), M. Mountain (STScI) and P. Puxley (NSF).*

■ *24 April 2006. A combination of X-ray, infrared and radio observations make up this image of the galaxy NGC 4696. A vast cloud of hot gas (red) surrounds high-energy bubbles 10,000 light-years across (blue) on either side of the bright white area around the central supermassive black hole. Data from this image led scientists to conclude that the conversion of energy by matter falling into the black hole is super-efficient. (The green dots in the image show infrared radiation from star clusters on the outer edges of the galaxy). Images courtesy NASA/CXC/KIPAC/S. Allen et al. – X-ray; NRAO/VLA/G. Taylor – Radio; NASA/ESA/McMaster Univ./W. Harris – Infrared.*

24 April 2006
Messier 82 helps Hubble celebrate 16 years in orbit

The European Space Agency and NASA celebrated the sixteenth birthday of the telescope by releasing a spectacular image of the starburst galaxy, Messier 82. M82 is a favorite of amateur astronomers, and its cigar-shaped smudge of light is visible in small telescopes. It is located 12 million light-years away in the constellation of Ursa Major.

M82 is a remarkable galaxy for a number of reasons. The star formation rate is ten times that found in our own galaxy, the Milky Way. Driven by huge stellar winds, plumes of hydrogen gas leap above and below the disk of the galaxy.

The image was taken in March 2006 by the Advanced Camera for Surveys. Like all color images from any professional telescope, images are composites of black and white images taken through specific color filters used to isolate the emission of different gases. The image showing ionized hydrogen gas, for example, is colored red, since the emission occurs in the red part of the spectrum. This image is a composite of six individual images extending from visible wavelengths to infrared. Infrared light is invisible to our eyes, so representative color used to illustrate those wavelengths.

http://www.spacetelescope.org/images/html/heic0604a.html
http://xxx.lanl.gov/abs/astro-ph/0612547

24 April 2006
Chandra Finds Black Holes Are "Green"

Black holes conserve energy very well. This is the conclusion of astronomers studying nine supermassive black holes with the Chandra X-ray Observatory. The black holes, found near the center of giant elliptical galaxies, each produce high energy jets traveling at near the speed of light in opposite directions from the black hole. Infalling matter provides the fuel for these jets, and astronomers wanted to find out just how energy efficient, or "green", the black holes were. The results show that the jets carry away most of the energy released as matter falls towards the black hole.

"If a car was as fuel-efficient as these black holes, it could theoretically travel over a billion miles on a gallon of gas," said team member Dr. Christopher Reynolds of the University of Maryland, College Park.

The black holes found in elliptical galaxies are far more quiescent than energetic quasars, yet more energy is being emitted by high energy particles than in visible light or by X-rays. The jets punch through the hot, X-ray emitting, intergalactic gas, creating large cavities or bubbles. The energy required to inflate these bubbles provides a perfect way to measure the power of the jets.

With such high energy being deposited into the galaxy, the hot gas is less likely to cool, becoming less inclined to collapse to form new sites of star formation. Consequently the black hole central engines generating these jets place an upper limit on the growth of the largest galaxies.

http://chandra.harvard.edu/photo/2006/bhcen/
http://xxx.lanl.gov/abs/astro-ph/0602549

■ 26 April 2006. The Spitzer Space Telescope captured this dramatic view of the colliding pair of galaxies, NGC 2207 and IC 2163. Giant beads showing large regions of active star formation are thought to have been caused by waves generated during the collision, creating the even pattern along the spiral arms. This picture, taken by Spitzer's infrared array camera, is a four-channel composite. It shows light with wavelengths of 3.6 microns (blue); 4.5 microns (green); and 5.8 and 8.0 microns (red). Starlight has been intentionally subtracted from the image to enhance the dusty features. Image courtesy NASA/JPL-Caltech/D. Elmegreen (Vassar).

26 April 2006

Galaxies Don Mask of Stars in New Spitzer Image

In an eye-catching image, the infrared Spitzer Space Telescope reveals a brilliant chain of star forming regions that appear to have been produced when two galaxies collided.

The brilliant central bulges of the two galaxies, NGC 2207 and IC 2163, are colored blue, resembling a couple of eyes peering through a delicate red mask of dust. The colors in the image are representative of different features, with gas showing as red and stars indicated by blue. The galaxies lie 140 million light-years from Earth in the constellation of Canis Major.

One spiral arm in particular appears to be studded with a regular pattern of bright concentrations. Similar concentrations dot the entire pair of galaxies. These are dusty clusters of newborn stars, made brightly visible by viewing them in infrared light.

"This is the most elaborate case of beading we've seen in galaxies," said Dr. Debra Elmegreen of Vassar College in Poughkeepsie, N.Y. "They are evenly spaced and sized along the arms of both galaxies."

The beads were presumably formed when the two galaxies first met. The gravitational perturbations set up waves of star formation along the spiral arms. A bright ring of star formation along oval arcs, with the appearance of an eyelid, is indicative of formations predicted to occur in grazing encounters of two galaxies. There are over 200 clumps along the spiral arms or the oval arcs.

The brightest clump is so large that it accounts for five percent of the total infrared light coming from both galaxies. Such a huge cluster could be so dense, astronomers suggest, that the central stars of the cluster may have merged to form a black hole.

http://www.spitzer.caltech.edu/Media/releases/ssc2006-11/release.shtml

http://xxx.lanl.gov/abs/astro-ph/0605524

http://xxx.lanl.gov/abs/astro-ph/0508660

MAY 2006

1 May 2006

VLBA Reveals Closest Pair of Supermassive Black Holes

The closest pair of supermassive black holes ever discovered has been found by astronomers using the Very Long Baseline Array (VLBA). The pair weighs in at more than 150 million times the mass of our Sun and lies at the center of the elliptical galaxy, 0402+379.

"These two giant black holes are only about 24 light-years apart, and that's more than 100 times closer than any pair found before," said Cristina Rodriguez of the University of New Mexico (UNM) and Simon Bolivar University in Venezuela.

The black holes orbit each other once every 150,000 years. Calculations show that the gravitational wave frequency emitted from this pair is far below current and near future detectable limits, and would appear in the noise level of the proposed LISA experiment.

However, Dr. Gregory Taylor of the University of New Mexico explains, "If two black holes like these were to collide, that event would create the type of strong gravitational waves that physicists hope to detect with instruments now under construction."

The pair of black holes is thought to have occurred through the process of a galaxy merger event. Such events are common over the history of the universe, and they have important effects on the subsequent evolution of the galaxies.

This new discovery represents one out of 293 sources surveyed so far. A new survey proposed by Taylor and colleague is already underway. The VLBA Imaging and Polarization survey (VIPS) includes 1,169 sources, raising the likelihood of finding even closer supermassive black hole binaries in that larger sample. (See also 6 April and 18 April stories.)

http://www.nrao.edu/pr/2006/binarybh/
http://xxx.lanl.gov/abs/astro-ph/0604042

3 May 2006

Companion Explains "Chameleon" Supernova

A supernova that exploded in 2001 and had mysteriously changed from a Type II to a Type I has finally revealed the reason why to observers using the Gemini South telescope. The supernova, SN 2001ig, showed signs of hydrogen in its spectrum, making it a Type II outburst. But later, the hydrogen lines disappeared, making it a Type I. How could this happen? It turns out that the star that exploded had a companion.

Astronomers using the Hubble Space Telescope have already faced a similar problem with SN1993J. They found, a whole decade after the explosion, the progenitor had a previously unseen companion star that had stripped off most of the outer hydrogen envelope. In such a system, after the star explodes, the hydrogen quickly disappears, becoming a Type I supernova.

A group of Australian astronomers used Gemini South telescope in Chile to detect the predicted companion star. The presence of the companion would explain why the supernova, which began looking like one kind of exploding star, seemed to change its identity after a few weeks.

Radio observations from the Australian Telescope Compact Array already hinted at a companion. They detected a clumpiness to the ejecta from the star. One explanation was that the ejecta was being disturbed into a spiral pattern by a companion star. Once the material cleared, the companion star should be visible.

The Gemini Multi-Object Spectrograph (GMOS) camera on the 8-meter Gemini South telescope swung into action and successfully detected a yellow-green point-like object at the location of the supernova explosion. It's the first time such a companion has been detected from the ground.

Dr. Stuart Ryder of the Anglo-Australian Observatory (AAO) notes, "We believe this is the companion. It's too red to be a patch of ionized hydrogen, and too blue to be part of the supernova remnant itself." Ryder and his team estimate the companion's mass to be between 10 and 18 times that of the Sun.

http://www.gemini.edu/
and click "Announcements Archive"

http://au.arxiv.org/abs/astro-ph/0603336

■ *Left: 1 May 2006. The Very Long Baseline Array (VLBA) spans more than 8,000 kilometers, from Hawaii to the US Virgin Islands. Ten radio telescope antennas are linked together during VLBA observations to simulate a dish 8,000 kilometers wide, producing the highest resolution of any telescope on Earth or in space. The VLBA was able to detect two supermassive black holes orbiting just 24 light-years apart at the center of an elliptical galaxy. Image courtesy NRAO/AUI/NSF.*

5 May 2006

New CARMA Array Brings Cool, Far Out Astrophysics More in Focus

The complex process of moving a network of telescopes to a new high altitude site in California reached a major milestone on 5 May as astronomers dedicated the new array. The Combined Array for Research in Millimeter-wave Astronomy, or CARMA, is a combination of the Berkeley-Illinois-Maryland Association (BIMA) millimeter array and the Owens Valley Radio Observatory (OVRO) millimeter array. It forms the largest millimeter-wavelength radio telescope in the world.

CARMA is a linked array of 15 radio telescope dishes located at 7,200 feet (2,195 meters) altitude in the dry desert of eastern

California's Inyo Mountains. Such an altitude places the radio telescopes above a significant portion of water vapor in Earth's atmosphere, a troubling source of interference for millimeter wavelengths. Elevating the telescopes from their valley locations to this new altitude is equivalent to an improvement in collecting area of 50-100 percent, all without building any new antennas.

"These observations will address some of the most important questions in astrophysics today," said Lewis Snyder, a professor emeritus of astronomy at the University of Illinois and a leader in the effort to develop the combined array. "These questions include how the first stars and galaxies formed, how stars and planetary systems like our own are formed, and what the chemistry of the interstellar gas can tell us about the origins of life."

The CARMA collaboration includes astronomers from the University of Maryland, the University of California at Berkeley, the University of Illinois at Urbana-Champaign and the California Institute of Technology.
http://www.mmarray.org/

■ *Above: 5 May 2006. One of six 10.4-meter antennas from the original Owens Valley Radio Telescope array is moved through a dramatic gorge on its way to join 14 other antennas to create the CARMA array, located in the high, dry desert of the Inyo Mountains in eastern California. Image courtesy Markus Raschke, University of Washington.*

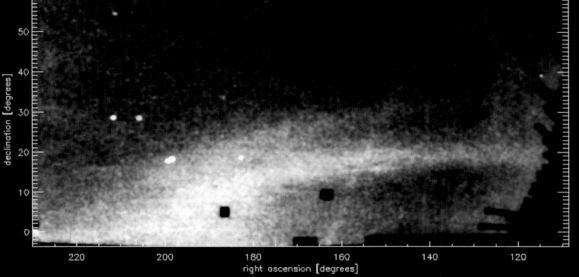

declination [degrees] / right ascension [degrees]

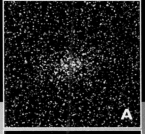

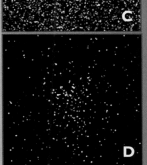

8 May 2006

Multiple Galaxy Mergers Continue in Milky Way

The Sloan Digital Sky Survey continues to produce ground-breaking results as more data is analyzed. Two results announced relate to our own Milky Way, revealing a far more complicated picture of our home galaxy than previously thought.

In the first result, newly found streams of stars criss-crossing large swaths of the northern sky represent the trails of satellite galaxies that have previously merged with the Milky Way. This so-called "Field of Streams" is the end result of dwarf galaxies that have been ripped apart by tidal forces of our own galaxy.

The second and related result is the announcement of two previously unknown satellite galaxies of the Milky Way, one in Canes Venatici and a second in Boötes. Both announcements were made by Cambridge University (UK) researchers Vasily Belokurov and Daniel Zucker.

Five years of data were analyzed to find the Field of Streams. The stars trace out the previous orbital path of the satellite galaxies, rather like meteor streams represent the orbital paths of comets within our solar system. The Sagittarius dwarf galaxy discovered over ten years ago dominates the stream, except the new survey revealed a split or fork in the stream, suggesting the stream goes around the Milky Way more than once.

In studying these streams, astronomers are becoming archeologists of the Milky Way, because they are digging up a chaotic and ancient history of mergers, and revealing those going on right now. The two new dwarf galaxies were discovered while studying the Field of Streams data.

"I was poring over the survey's map of distant stars in the Northern Galactic sky – what we call a Field of Streams – and noticed an overdensity in Canes Venatici," Zucker explained. "Looking further, it proved to be a previously unknown dwarf galaxy. It's about 640,000 light-years (200 kiloparsecs) from the Sun. This makes it one of the most remote of the Milky Way's companion galaxies."

Within hours of Zucker's discovery, Belokurov had found his own dwarf galaxy in Boötes. Its distorted structure hints at it being disrupted by the Milky Way's gravitational tides.

Mark Wilkinson, a collaborator on the science team, places these discoveries in context. "Finding and studying these small galaxies is really important. From their structure and their motions, we can learn about the properties of dark matter, as well as measure the mass and the gravity field of the Milky Way," he said.

http://xxx.lanl.gov/abs/astro-ph/0605705

http://xxx.lanl.gov/abs/astro-ph/0604355

11 May 2006

XMM-Newton Reveals the Origin of Elements in Galaxy Clusters

Fundamental to our very existence are the presence of chemical elements. One of the basic questions is: where do the elements come from? It's a great success of modern science that the origin of the elements is generally understood. A mixture of hydrogen and helium, plus some traces of light elements, were produced by nucleosynthesis during the Big Bang before the temperature and pressure dropped too low. The remaining chemical elements (the vast majority) come from supernova explosions.

So it should come as no surprise that the precise determination of chemical abundances in clusters of distant galaxies is crucial to understanding the evolution of chemicals in the universe. Helping to trace the origin of different elements is the fact that the two main types of supernova, Type Ia and Type II, inject different proportions of elements into the intergalactic medium. The core collapsing Type IIs generate a lot of oxygen, neon and magnesium, whereas the disintegration of a white dwarf in a Type Ia supernova generates iron and nickel.

The signature of many chemical elements, including the first detection of chromium, has recently been measured to high accuracy by the European Space Agency's large X-ray observatory, XMM-Newton. Observing two X-ray bright galaxy clusters, Newton found the hot X-ray gas weighs in as much as five times the mass of the galaxies themselves, and is populated by gas from supernova explosions and fast stellar winds from very hot stars. Both clusters naturally exhibit a range of products from both types of supernovas, offering the opportunity to determine a ratio of each type of explosion in each cluster.

Norbert Werner from the SRON Netherlands Institute for Space Research explains the results. "Comparing the abundances of the detected elements to the yields of supernovae calculated theoretically, we found that about 30 percent of the supernovae in these clusters were exploding white dwarfs (Type Ia) and the rest were collapsing stars at the end of their lives (core collapse)," he said

This number fits well within the lower number of 13 percent of Type Ia supernovas found in our galaxy, and 42 percent determined by the Lick Observatory Supernova Search Project.

http://sci.esa.int/xmm/
http://xxx.lanl.gov/abs/astro-ph/0512401
http://xxx.lanl.gov/abs/astro-ph/0602582

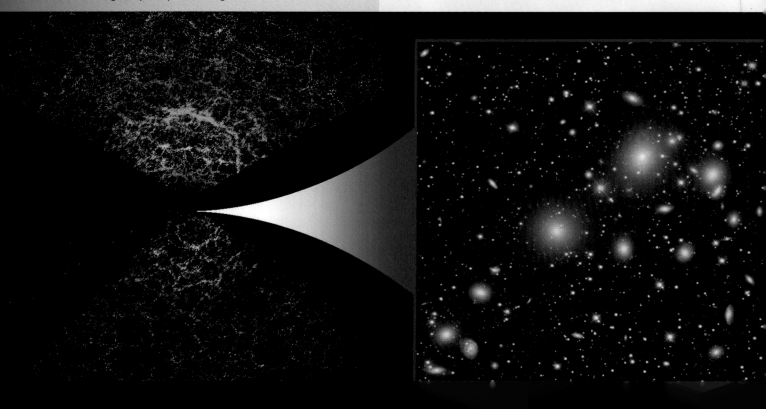

■ *Below: 15 May 2006. The Sloan Digital Sky Survey combines images (right) with the distances to each galaxy determined from their spectrum to create a three-dimensional map of the universe. This example shows a map spanning 2 billion light-years. Each galaxy is shown as a single point, with the color representing the luminosity. This map shows 66,976 out of 205,443 galaxies in the map that lie near the plane of Earth's equator. The new survey announced on 15 May contains a million galaxies spanning 5 billion light-years. Image courtesy Sloan Digital Sky Survey and Max Tegmark.*

15 May 2006

Astronomers Construct Large 3D Map of a Million Galaxies

The sheer volume of data collected by sky surveys such as the Sloan Digital Sky Survey (SDSS) is making life difficult for astronomers trying to determine accurate distances to far off galaxies. There are just so many galaxies. Distance is determined with Hubble's Law by combining the measured redshift of each galaxy with the current value of the Hubble constant. Given a million galaxies, it could take more than a lifetime to create a 3D map without some smart tools.

Such a problem had a solution to Professor Ofer Lahav of University College, London, England and Dr. Adrian Collister of Cambridge University, England. They developed an artificial neural-network-based computer algorithm which could identify Luminous Red Galaxies (LRG) in the SDSS. The painstaking work was in determining accurate distances to 13,000 galaxies in order to "train" the algorithm to do the correct calculation of redshifts. A survey called 2SLAQ, part of the 2-Degree Field survey of LRG's and Quasars undertaken by an international team of astronomers using the Anglo Australian Observatory, provided a suitable data set.

By applying their newly-trained software algorithm to the entire SDSS, Lahav and Collister came up with over a million galaxies spanning 5 billion light-years, allowing creation of the MegaZ-LRG catalog.

Analysis of the huge catalog undertaken by Dr Chris Blake of the University of British Columbia, Canada, confirms some of the cosmological parameters determined by the Wilkinson Microwave Anisotropy Probe mission, such as the idea that we live in a universe dominated by dark energy and dark matter. The echo of sound waves in the early universe is imprinted on the large scale structure of the relatively nearby universe.

"We have analyzed the patterns in this map and discovered waves of structure over a billion light-years across," said Chris Blake. "These waves were generated billions of years ago and have been vastly stretched in size by the expanding Universe."

In a parallel and independent survey by Nikhil Padmanabhan of Princeton University analyzing a similar set of data, similar results were obtained.

"By comparing the new measurements to the microwave background data, astronomers can test whether these enormous cosmic structures have grown at the expected rate – between the time the cosmic microwaves were emitted and the time that the light of the new structures was emitted," said Padmanabhan. Princeton colleague Uros Seljak adds, "With the new measurements, our emerging picture of a Universe dominated by dark matter and dark energy had a chance to fall on its face. Instead, it passed the new test with flying colors".

http://www.sdss.org
http://www.2slaq.info/
http://xxx.lanl.gov/abs/astro-ph/0607630
http://xxx.lanl.gov/abs/astro-ph/0605303
http://xxx.lanl.gov/abs/astro-ph/0607631
http://xxx.lanl.gov/abs/astro-ph/0605302

17 May 2006

Site in Northern Chile Selected for Large Synoptic Survey Telescope

Following a two-year long intensive study of four sites distributed around the entire globe, the location for the proposed Large Synoptic Survey Telescope (LSST) has been chosen. The 8.4-

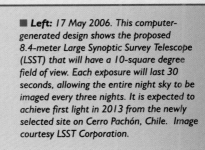

■ **Left:** 17 May 2006. This computer-generated design shows the proposed 8.4-meter Large Synoptic Survey Telescope (LSST) that will have a 10-square degree field of view. Each exposure will last 30 seconds, allowing the entire night sky to be imaged every three nights. It is expected to achieve first light in 2013 from the newly selected site on Cerro Pachón, Chile. Image courtesy LSST Corporation.

3 μm 4 μm 7 μm

11 μm 15 μm 24 μm

meter wide-field telescope will be built on Cerro Pachón, an 8,800-foot (2,682-meter) mountain peak in northern Chile.

LSST will survey the entire visible sky every three nights using a massive 3 billion pixel digital camera. The telescope, once operational, will change how astronomers, both professional and amateur, search for comets, supernovas, variable stars, and Near Earth Objects.

Taking images of the sky on such a large scale will enable subtle effects of gravitational lensing to be revealed, in addition to probing the mysteries of dark matter and dark energy. The telescope is expected to achieve "first light" in 2012.

http://www.lsst.org/News/
http://xxx.lanl.gov/abs/astro-ph/0303012
http://xxx.lanl.gov/abs/astro-ph/0701506

■ *Above: 22 May 2006. The Japanese AKARI infrared observatory took these images of the galaxy M81 using the near- and mid-Infrared Camera (IRC). The different wavelengths detect different aspects of the galaxy's content. The shorter wavelengths (3 and 4 microns) reveal stars free from obscuring dust. The intermediate wavebands centered at 7 and 11 microns detects radiation from organic materials in the interstellar gas. The two longest wavelengths reveal dust heated by young hot stars (15 and 24 microns). M81 is a spiral galaxy located at a distance of about 12 million light-years from us.*
Image courtesy JAXA.

22 May 2006

Infrared Space Observatory "Akari" Delivers Its First Images

Launched on 21 February 2006, the Japanese infrared space observatory, Akari, returned its first outstanding images following completion of a two month check-out procedure. It achieved first light on 13 April 2006.

The marvelous images reveal the greater sensitivity and resolution over the previous all-sky survey mission, IRAS. The IRAS survey remains in extensive use today, and the Akari survey promises to be even more successful.

Two instruments, the Far Infrared Surveyor (FIS) and the near-mid-infrared camera (IRC), are undertaking an all-sky survey in six infrared wavebands. It will take six months to complete. Detailed targeted observations of specific objects will follow.

The mission has some support from the European Space Agency, which is actively developing its own plans for Herschel, a 3.5-meter deployable infrared space telescope planned for a 2007 launch, that will be able to follow up on interesting objects found in the Akari survey.

http://www.jaxa.jp/projects/sat/astro_f/index_e.html

http://www.esa.int/esaSC/SEM8NF9ATME_index_0.html

23 May 2006

Hubble Captures a 'Five-Star' Rated Gravitational Lens

The first-ever picture of a quasar lensed by gravity into five separate images has been obtained by the Hubble Space Telescope. The remarkable image shows a huge cluster of galaxies, one of the most distant known at 7 billion light-years away. The cluster lies in the foreground compared to the 10 billion light-year distance of the quasar. The quasar's light is bent by the space-warping gravity of the huge agglomeration of galaxies. The amount of bending allows astronomers to calculate the mass of the intervening cluster.

How do astronomers know there are not five separate quasars that happen to lie close together on the sky? When the detailed spectra of each quasar image, taken by the Keck telescope, were analyzed, astronomers found them to be identical. The likelihood that five quasars with identical spectra all lie in such a small area of sky is very small. Also other lensed images around the cluster, such as the galaxy arcs, are visible in the image, making the lensing of a single quasar far more likely.

Four of the quasar images are clearly visible, and have actually been seen before, but the fifth is superimposed close to the central galaxy. Hubble's original image enables the core of the galaxy to be seen separate from the quasar. The host galaxy of the quasar is visible as a red arc associated with each quasar image. Gravitational lensing always produces an odd number of images.

http://www.spacetelescope.org/news/html/heic0606.html

http://hubblesite.org/newscenter/archive/releases/2006/2006/23/

■ **Below:** *23 May 2006. The first-ever picture of a group of five star-like images of a single quasar taken by the Hubble Space Telescope. The multiple images are created by a huge gravitational lens created by an intervening cluster of galaxies. The cluster, found by the Sloan Digital Sky Survey, lies 7 billion light-years away. The quasar lies 10 billion light-years away. A spectrum of each of the star-like images by the 10-meter Keck telescope confirmed they are one and the same quasar. Four of the images are clearly seen, and the fifth lies very close to the central galaxy. Image courtesy ESA, NASA, K. Sharon (Tel Aviv University) and E. Ofek (Caltech).*

JUNE 2006

1 June 2006

Thirty Meter Telescope Passes Conceptual Design Review

The Thirty Meter Telescope proceeds to a detailed design stage following successful completion of its conceptual design review by an independent panel. Pushing the limits of engineering design, the TMT will carry a mind-boggling 738 individual 1.2 meter mirror segments spanning 30 meters, offering nearly ten times the light collecting power of existing telescopes. The plan is for first light to occur in early 2016, three years after the expected launch of the James Webb Space Telescope (JWST). Astronomers hope that the TMT will complement JWST in the same way the 10-meter Keck telescopes complement Hubble Space Telescope observations.

The TMT project received the highest priority for ground-based astronomy in the National Academy of Sciences decadel survey published in 2000. The projected cost of the design and development phase is $64 million, nearly half of which will come from a private donation by the Gordon and Betty Moore Foundation.

TMT will work primarily in the infrared region of the spectrum, and be a frontline tool in the search for the first stars and galaxies that formed after the Big Bang. The telescope plans to use the adaptive optics technique pioneered by Lick Observatory, and by the Gemini and Keck telescopes.

The decision for the site of the telescope will be made in 2008, followed by construction which is planned to start in 2009. Five sites are under review, in Chile, Hawaii and Mexico.

http://www.tmt.org/

The following news stories from 5-9 June 2006 were presented at the Calgary meeting of the American Astronomical Society.

5 June 2006

"Cosmic Telescopes" May Have Found Infant Galaxies

The search for infant galaxies in the early universe is one of holy grails of astronomy. Valuable tools in this search are giant clusters of galaxies. Their huge gravitational influences act like giant telescopes, bending and focusing light rays of much more distant galaxies.

"One of Einstein's most startling predictions was that a gravitational field can be thought of as a distortion of space and time," noted Holland Ford, the head of the Hubble Advanced Camera for Surveys at Johns Hopkins University. "Gravitational lensing by massive clusters of galaxies that have about 1 million billion times

■ *1 June 2006. The proposed Thirty Meter Telescope (TMT) is shown inside a cutaway of the dome. The huge telescope will contain more than 700 hexagonal-shaped mirrors that span 30-meters across. The National Academy of Sciences report "Astronomy and Astrophysics in the New Millennium" made the TMT its highest-priority for new ground-based facilities for the first decade of the 21st century. The design concept recently passed a major review. First light is expected in 2016. Image courtesy Thirty Meter Telescope Project.*

more mass than the Sun provides one of the most striking confirmations of Einstein's prediction," he added.

By using such clusters, astronomers are able to see galaxies born in the first billion years after the Big Bang. The lensing can magnify such clusters 50 to 100 times. The technique compares optical images taken with the Hubble Space Telescope's Advanced Camera for Surveys with infrared images taken with the giant 8-meter Gemini and 10-meter Keck telescopes. The infant galaxies are so far away their light has redshifted to infrared wavelengths beyond the reach of Hubble's camera, but within reach of the larger ground-based telescopes.

The science team has selected 14 clusters to search, and if any of the galaxies detected as lensed objects on the ground show spectroscopic features of highly redshifted galaxies that are in the early universe, the team will have found their quarry.

http://www.jhu.edu/news_info/news/home06/jun06/cosmic.html

5 June 2006

Planetary-Mass Objects Found to be Surrounded by Discs

The first dusty disk around another star was dramatic news in 1983, when Beta Pictoris was discovered to possess one. Since then many stars have revealed their dusty disks, believed to be the precursor to, or in some cases the remnants of, planetary formation around other stars. However, apparently disks are not restricted to stars. In recent years, evidence for dusty disks around failed stars called brown dwarfs has been obtained.

Now the same astronomers who found the brown dwarf disks have gone one step further, and discovered disks around planetary mass objects, or planemos. Two sets of observations led by Professor Ray Jayawardhana of the University of Toronto, and involving astronomers from the USA, Europe and Chile, made the discoveries. The planemos, weighing in a few times the mass of Jupiter, and in some cases existing on their own and not orbiting a star, show infrared emission from dusty disks that may, some time in the future, evolve into moons orbiting the planetary objects.

One target, located 170 light-years away in the constellation of Centaurus, is a brown dwarf of 25 times Jupiter's mass orbited by an 8-Jupiter mass planemo in an orbit equivalent to Pluto around our Sun. While the brown dwarf has its own disk, it turns out the planemo does also.

In another observation, four objects in a star forming region 450 light-years away and weighing between 5 and 15 times the mass of Jupiter each possess a dusty disk. The observations were made by the European Southern Observatory's 8.2-meter Very Large Telescope and the 3.5-meter New Technology Telescope.

"Now that we know of these planetary mass objects with their own little infant planetary systems, the definition of the word 'planet' has blurred even more," says Ray Jayawardhana. "In a way, the new discoveries are not too surprising - after all, Jupiter must have been born with its own disk, out of which its bigger moons formed," he said.

http://www.cfa.harvard.edu/press/pr0616image.html
http://arxiv.org/abs/astro-ph/0610550
http://arxiv.org/abs/astro-ph/0607152

5 June 2006

Andromeda Galaxy I: Spitzer Space Telescope Views M31

In the first of five featured research studies of the giant Andromeda galaxy (M31), take a look at our neighboring large spiral galaxy from the big picture, wide-angle, view provided by the Spitzer Space Telescope (SST). Later, follow the story by zooming into its heart with the 8.2-meter Gemini telescope, see what the X-ray's reveal, and trace carbon monoxide throughout the galaxy.

The SST has obtained remarkably detailed images of the entire Andromeda galaxy in a montage that traces star forming regions all the way to the inner region of the galaxy, and shows for the first time evidence of a collision with a nearby companion. The galaxy spans more than seven Moon-widths in the sky, and required more than 3,000 individual images from Spitzer to create the finished image. M31 spans 260,000 light-years from edge to edge, compared to the Milky Way's 100,000 light-years.

The longer infrared waveband (shown as red) reveals dust throughout the galaxy, regions that house active star formation, and a shorter waveband (shown as blue) reveals old stars. For the first time, astronomers used the infrared luminosity of the galaxy to estimate the total number of stars in M31. There are about one trillion stars in the galaxy, compared to about two hundred billion in our own Milky Way, confirming that M31 is significantly larger than our galaxy. The Andromeda galaxy lies about 2.5 million light-years from Earth.

"This is the first time the stellar population of Andromeda has been determined using the galaxy's infrared brightness," said Dr. Pauline Barmby of the Harvard-Smithsonian Center for Astrophysics, Cambridge, Mass. "It's reassuring to know our numbers are in agreement with previous estimates of the mass of the stars based on the stars' motion."

The image also reveals a stark contrast between the smooth disk of stars and the chaotic appearance of its dusty regions. In research by Barmby and other colleagues published later in October 2006 (see astro-ph link below), a new asymmetric inner ring of dust is seen in the SST images. Rings are formed either by collision or by rotating bars. Scientists conclude that M31 experienced a head-on collision with a companion galaxy 210 million years ago, most likely with the dwarf galaxy, M32.

http://www.spitzer.caltech.edu/Media/releases/ssc2006-14/

http://arxiv.org/abs/astro-ph/0608593

http://arxiv.org/abs/astro-ph/0610543

5 June 2006

Andromeda Galaxy II: Gemini Peers into the Heart of Andromeda

The 8.2-meter Gemini North telescope and its high resolution adaptive optics, combined with a sensitive infrared camera, has probed deep into the heart of the Andromeda galaxy, revealing details far beyond the reach of the 0.85-meter Spitzer Space Telescope. The results reveal thousands of individual stars within 6,500 light-years of the core of M31 .

Knut Olsen of the National Optical Astronomy Observatory's and his team (including Stephen Strom, whose article on star formation appears in Chapter 5 of this book) compared the distribution of the brightness of individual stars with our knowledge of how the luminosities of stars evolve over time. They show the disk of M31 has been around for about half the age of the universe, or 6 billion years. Most of the stars are relatively old and have heavy element compositions similar to the Sun. According to current theories of galaxy formation, galaxies grow my gentle merging of clumps because violent merges tend to destroy galaxy disks. The population of stars found in M31 offers evidence of when merger activity abated.

http://www.gemini.edu/aasm31
http://arxiv.org/abs/astro-ph/0603793

5 June 2006

Andromeda Galaxy III: To the Core of Andromeda

In addition to the thousands of stars found in the core of M31, one intriguing dust-enshrouded star lies near its center. Dr. Tim Davidge of the Herzberg Institute of Astrophysics, Victoria, BC, Canada and his team has focused on this object that reveals some similarities with the core of our own galaxy.

The core is known to contain a double nucleus, one of which harbors a supermassive black hole. Because of the shredding effect of the intense gravitational field created by the black hole, clouds of gas and dust were thought to be rare, with the core populated only by old stars.

New images by Gemini reveal a highly-evolved massive star known to throw copious amounts of dust into space. This so-called "asymptotic giant branch" star (referring to a section of the Hertzsprung-Russell diagram in which it lies) is relatively short-lived compared to other stars in M31's core, suggesting it's a relatively young object. It lies between the galaxy's two nuclei, is estimated to be about a few hundred million years old, and is similar to stars found in the core of the Milky Way.

"Now we see that the centers of M31 and the Milky Way may be more similar than once thought," said Davidge. "These two neighbors in space share some similarities, although not where we might expect. This agreement gives us hope that the center of the Milky Way may be representative of other galaxies. If so we can use our home galaxy as a laboratory to understand much more distant galaxies."

http://www.gemini.edu/aasm31
http://arxiv.org/abs/astro-ph/0605006

Stars & Dust

Stars

Dust

■ **Left:** *5 June 2006. The central part of the Spitzer Space Telescope infrared image of M31 is combined with the view in X-rays from the Chandra X-ray Observatory, two of NASA's Great Observatories program. The infrared view is shown red, low energy X-rays are green and high energy X-rays are blue. The view spans an area about the size of a full Moon. The diffuse glow from X-rays is due to gas heated to millions of degrees, possibly by shock waves from supernova explosions in the busy central regions of M31. Images courtesy NASA/UMass/Z. Li & Q.D. Wang – X-ray; NASA/JPL-Caltech – Infrared.*

5 June 2006

Andromeda Galaxy IV: Chandra X-ray Observatory Views Andromeda

Imaging M31 at X-ray wavelengths provides a completely different view of our neighboring spiral galaxy compared to the infrared view. The image is dominated with point sources that are likely binary star systems in which one companion is a neutron star or a black hole. Matter is being pulled from its companion and heated to tens of millions of degrees, producing X-rays.

The central region reveals a diffuse X-ray emitting cloud of hot gas that is centered on the core of the galaxy. Heating of this galactic gas is thought to be powered by supernova explosions. This energy could also drive gas away from the center of the galaxy, depleting raw materials for star formation and preventing further infall of gas into the core. The point sources detected in X-rays are probably binary star systems in which one component is a neutron star or black hole. Their strong gravity draws material off the companion, forming an accretion disk. As additional material collides with the disk, X-rays are generated.

http://chandra.harvard.edu

http://arxiv.org/abs/astro-ph/0408305

■ *7 June 2006. The galaxy NGC 5866 appears practically edge-on to our line of sight. A clear dust lane divides the galaxy into two halves. The galaxy, if viewed face-on, would look like a smooth flat disk with few spiral arms. A blue disk of hot young stars runs parallel to the dust lane. The large central bulge is reminiscent of an elliptical galaxy. This is an example of a lenticular galaxy, and lies in the direction of the constellation Draco, 44 million light-years away. Image courtesy NASA, ESA, and The Hubble Heritage Team (STScI/AURA).*

5 June 2006

Most Distant Cluster Of Galaxies Found by XMM-Newton

Astronomers announced the discovery of a cluster of galaxies 10 billion light-years from Earth. The cluster, called XMMXCS 2215-1738, was found from its hot- X-ray emitting gas and was detected by the European Space Agency's XMM-Newton X-Ray observatory. The 10-meter Keck telescope determined its distance. However, the new cluster presents a challenge to astronomers, since it contains apparently old galaxies when the universe was just one quarter of its present 13.7 billion year age.

"This cluster is a challenge for our models of how massive galaxies formed, and to our understanding of the way such a massive cluster exists at a relatively early era in the Universe, said Adam Stanford from UC Davis and the Lawrence Livermore National Laboratory.

Using the temperature of the X-ray emitting gas, astronomers determined the mass of stars in the cluster to be 500 trillion times the mass of the Sun. Such early formation of massive galaxies is only expected in models that incorporate dark energy, according to cosmologist Dr Pedro Viana from the University of Porto, Portugal. "It is yet more evidence that we live in a strange Universe," he said.

The new results came from the X-ray Cluster Survey (XCS) currently being performed by XMM-Newton. Already scientists have data from thousands of other sources and plan to determine their distances. This record-breaker was found on the scientist's first night at Keck. "I can't wait to find out how many more clusters like XMMXCS 2215-1738 there are out there," said team member Dr Kathy Romer of the University of Sussex, England.

http://xcs-home.org/aas

http://arxiv.org/abs/astro-ph/0606075

7 June 2006

Hubble Views Edge-on Galaxy

The galaxy NGC 5866 lies practically edge-on to our line of site, offering the Hubble Space Telescope a clear view of its dust lane representative of a spiral galaxy. This galaxy has a fairly smooth disk, and if viewed face on would show few spiral arms, and is classed as an S0 spiral. A blue disk parallels the dust revealing light from hot young O and B-type stars. The dust disk is slightly warped, suggesting a recent tidal disturbance by a near encounter with another galaxy. NGC 5866 is a member of a small cluster of galaxies in which near encounters are likely over its lifetime.

NGC 5866 lies 44 million light-years away in the direction of the constellation of Draco, and spans about two-thirds of the diameter of our Milky Way, at about 60,000 light-years.

http://hubblesite.org/newscenter/archive/releases/2006/24

19 June 2006

Full Speed Ahead for Cosmic Ray Project

The most powerful cosmic ray detector in the northern hemisphere got a green light for its final stages of construction in Utah this month, following a building permit from the Bureau of Land Management, and a $2.4 million grant from the National Science Foundation.

Cosmic rays have remained a mystery every since their discovery by Victor Hess in 1912. The high energy atomic nuclei strike nitrogen in the upper atmosphere and create air showers of subatomic particles. Some originate in supernova explosions, but others are suspected to come from the far reaches of the universe, such as the cores of active galactic nuclei, which contain supermassive black holes, or even from the Big Bang itself.

The Telescope Array has two components. Three "fluorescence detectors" record faint ultraviolet flashes caused by cosmic rays hitting the atmosphere. The resulting Cerenkov radiation can be triangulated using the three stations, and the cosmic ray's original direction determined. The second component is an 29-kilometer by 35-kilometer wide ground array of 564 scintillation detectors that will record air showers of subatomic particles caused by the original cosmic ray impact with nitrogen in our atmosphere.

Professor Pierre Sokolsky, chair of physics at the University of Utah, expects the new observatory to begin test runs in late spring 2007, followed by full operations in the late summer.

The Telescope Array will be 10 times more sensitive than previous experiments, and we hope it will allow us to finally resolve the mystery of the origin of these ultrahigh-energy particles [cosmic rays] that are bombarding the Earth," says Sokolsky.

The Telescope Array is a collaborative project between institutions in Japan, Taiwan, China and the United States.

http://www.telescopearray.org/index.html

27 June 2006

Hubble Reveals Two Dust Disks Around Beta Pictoris

Beta Pictoris has been known to sport a dusty disk since the Infrared Astronomy Satellite first detected it in 1983, and Brad Smith and Richard Terrile acquired the first ground based images of the disk the following year. A new image from the Hubble Space Telescope's Advanced Camera for Surveys (ACS) now reveals two distinct disks. Using the ACS coronograph, the new image shows a second disk tilted by about 5 degrees to the main disk, visible out to about 38 million kilometers from the star.

"The Hubble observation shows that it is not simply a warp in the dust disk but two concentrations of dust in two separate disks," said David Golimowski of Johns Hopkins University in Baltimore, Maryland. "The finding suggests that planets could be forming in two different planes. We know this can happen because the planets in our solar system are typically inclined to Earth's orbit by several degrees," he added.

The second disk implies at least one Jupiter-size planet orbiting the star, according to computer models performed by David Mouillet and Jean-Charles Augereau of Grenoble Observatory in France. If a massive planet is in orbit around the star, it attracts smaller rocky and icy objects into orbits aligned with the planet's orbit. Collisions of this smaller debris could create the observed disk.

"The actual lifetime of a dust grain is relatively short, maybe a few hundred thousand years," added Golimowski. "So the fact that we can still see these disks around a 10- to 20-million-year-old star means that the dust is being replenished by collisions between planetesimals," he said.

http://hubblesite.org/news/2006/25
http://arxiv.org/abs/astro-ph/0602292

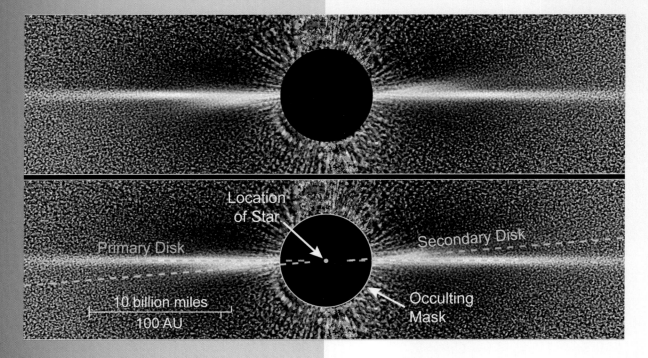

Location of Star

Primary Disk

Secondary Disk

10 billion miles
100 AU

Occulting Mask

■ **Above:** *27 June 2006. The Hubble Space Telescope acquired this new view of Beta Pictoris, revealing the main disk and a new secondary disk inclined 4 to 5 degrees. The new disk, revealed for the first time by the Advanced Camera for Surveys in coronagraph mode, is circumstantial evidence for planets in an inclined orbit. Image courtesy NASA, ESA, D. Golimowski (Johns Hopkins University), D. Ardila (IPAC), J. Krist (JPL), M. Clampin (GSFC), H. Ford (JHU), and G. Illingworth (UCO/Lick) and the ACS Science Team.*

■ **Below:** *29 June 2006. A new map of cold carbon monoxide in the Andromeda Galaxy (M31) was made by observations at millimeter wavelengths. The six year project took 800 hours of telescope time using the 30-meter Millimeter-wave telescope of the Institute for Radio Astronomy in the Millimeter range (IRAM). Carbon Monoxide is found deep within cold dust clouds near spiral arms, and represents future sources of star formation.*
Image courtesy Max-Planck-Institut für Radio Astronomie.

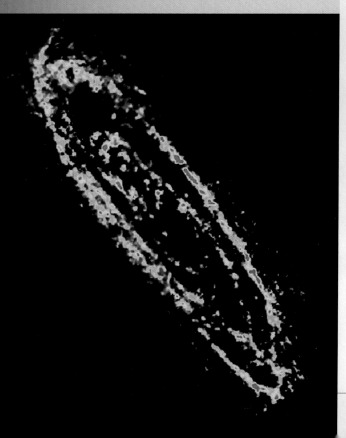

29 June 2006

Andromeda Galaxy V: Cold Gas in the Andromeda Galaxy

Completing a detailed look at the Andromeda galaxy this month, a new map of cold carbon monoxide in M31 has been completed after a mammoth effort. The project took six years, ending in 2001, and consumed about 800 hours of telescope time. The observations were made by astronomers from the Institute for Radio Astronomy in the Millimeter range (IRAM) in Grenoble, France, and Germany's Max-Planck Institute for Radio Astronomy. IRAM's 30 meter millimeter telescope, located on Pico Velata near Granada, Spain, made the observations.

Carbon monoxide (CO) traces clouds of dense, cold gas within the Andromeda galaxy, locations where new stars are born. CO is always in the presence of the hydrogen molecule whose radiation is essentially invisible from the Earth's surface. Therefore astronomers use CO as a tracer of cold hydrogen, and therefore locations in a galaxy of future star formation.

By measuring the intensity of the CO signal at many locations, a map of its distribution is constructed. Astronomers also measured any Doppler shift in the line from its rest wavelength, and were able to show that Andromeda is rotating at 200 kilometers per second about a central axis. Andromeda is also moving towards the Milky Way at 300 kilometers per second, and will pass close by in about 2 billion years.

The new map, recently released after extensive processing required to create it, shows the spiral arms at distances between 25,000 and 40,000 light-years from the centre of Andromeda, where most of the star formation occurs.

http://www.mpifr-bonn.mpg.de/public/pr/pr-m31-en.html
http://arxiv.org/abs/astro-ph/0512563

JULY 2006

1 July 2006

New Radio Telescope to Probe Early Universe

A simple arrangement of 8,000 dipole antennas lies at the heart of a new telescope probing the low frequency (80 to 300 Mhz) end of the electromagnetic spectrum. A $4.9 million grant from the National Science Foundation moves the Mileura Widefield Array (MWA) - Low Frequency Demonstrator (LFD) closer to full scale construction. The group is led by the Haystack Observatory and Massachusetts Institute of Technology (MIT), and CSIRO, the Australian scientific research agency. The new telescope is being built in a radio quiet zone in Western Australia.

The LFD will be spread over an area almost 1.5 kilometers in diameter. Because the telescope operates at the same frequencies as FM radio and television broadcasts, low frequency astronomy is best performed from remote sites, such as Western Australia.

In addition to monitoring nearby solar outbursts, astronomers will be using the LFD to attempt to detect very distant signals at the remote limit of the observable universe. When the first stars and galaxies formed, they created a pattern in the primordial hydrogen formed after the Big Bang and should be detectable in the low frequency domain.

"Radio astronomical telescopes operating at low frequency provide an opportunity to witness the formation of the first stars, galaxies and clusters of galaxies, and to test our theories of the origin of structure," said Jacqueline Hewitt, director of the MIT Kavli Institute

"Another exciting possibility also exists," explains University of Melbourne's Professor Rachel Webster. "We also hope to see spherical holes created by early quasars (active cores of galaxies) in the smooth distribution of primordial hydrogen. These will appear as small dark spots where the quasar radiation has split the hydrogen apart into protons and electrons," she said.

http://www.haystack.mit.edu/ast/arrays/mwa/
http://arxiv.org/abs/astro-ph/0611751

(Note: See also the feature article, High-stakes Astronomy at Low Frequencies....Opening Up a New Wavelength Frontier by Carolyn Collins Petersen in Chapter 4 of this book for more background on this new and exciting field of research.)

5 July 2006

A New Dome on Haleakala

A new 1.8-meter telescope called PS-1 was dedicated on 30 June, ushering in an exciting new prototype for the larger proposed Panoramic Survey Telescope and Rapid Response System (Pan-STARRS). The main job of the PS1 instrument is to survey the entire sky visible from Hawaii every few days, searching for asteroids, Near Earth Objects, and monitor other changes in the sky. The prototype program is planned to last 3.5 years.

The PS-1 telescope is located on the 10,000-foot (3,048-meter) summit of Haleakala, Hawaii. While the telescope's size is nothing remarkable, it's camera will be the world's largest digital camera, containing 1.4 billion pixels. While surveying the skies, PS1 will generate about 2,000 gigabytes of data, which will be later analyzed by the Maui High Performance Computing Center in Kihei, Hawai. The future Pan-STARRS telescope is scheduled to start scanning the skies for "killer asteroids" in 2010.

http://www.ifa.hawaii.edu/info/press-releases/PS1_dedication/

6 July 2006

Supernova Leaves Behind Mysterious Object

Conventional wisdom often expects a pulsar, a rapidly spinning neutron star, to be the aftermath of a supernova explosion, along with rapidly expanding ejecta. The supernova remnant RCW103 appeared to fit the bill, revealing the aftermath of a supernova that occurred nearly 2,000 years ago. However, a mystery surrounds the pulsar.

Young supernova remnants, and this one is young, are expected to have pulsars with periods of seconds, or milliseconds. New observations by the European XMM-Newton X-ray telescope have created quite a stir, because the periodicity found in emissions from RCW103 has a 6.7 hour period, thousands of times longer than scientists expected to find.

"The behaviour we see is especially puzzling in view of its young age, less than 2,000 years," said Andrea De Luca of IASF-INAF, the lead author. "It is reminiscent of a multimillion-year-old source. For years we have had a sense that the object is different, but we never knew how different until now."

If the period of variation is truly the rotation period of the central object, then a peculiar kind of neutron star called a magnetar could explain the slow period. Scientists suspect the object is one of two objects: an X-ray binary with a low mass star in an eccentric orbit around a compact object, or an isolated neutron star with a strong magnetic field. While a strong magnetic field could slow the pulsar down, it could not reach a 6.7 hour period in 2,000 years without additional help. If a debris disk formed from infalling material, this might also help slow the pulsar down. This would represent a new evolutionary model of neutron star

http://www.esa.int/esaCP/
SEM30QVT0PE_index_1.html
http://arxiv.org/abs/astro-ph/0607173

■ **Above:** 5 July 2006. A new 1.8-meter telescope called PS-1 is visible through the new dome of the telescope, located on Haleakala, Hawaii. The telescope is the prototype for a larger Panoramic Survey Telescope and Rapid Response System, or Pan-STARRS. The larger telescope's main task is to scan the skies for potentially hazardous asteroids that could threaten Earth. The PS-1 telescope will scan the skies every few days monitoring for changes. Image courtesy IFA, University of Hawaii.

■ **Right:** 6 July 2006. The 2,000 year old supernova remnant, RCW103, is shown in an image from the European Space Agency's XMM-Newton Observatory. The blue dot in the center is an unusual X-ray source. As the light curve shows to the right, the central object has a period of 6.7 hours. This is remarkably long for a young pulsar, and is a challenge for scientists to explain. Image courtesy ESA/XMM-Newton/A. De Luca (INAF-IASF).

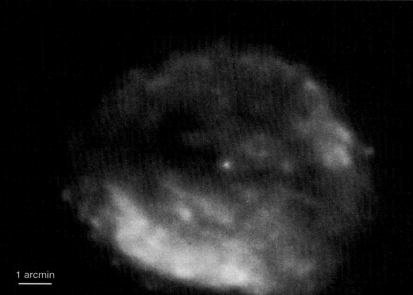

1 arcmin

1 - A Year in News and Pictures

19 July

Blast Teaching Astronomers New Lessons About Cosmic Explosions

White dwarfs are the progenitors of a class of supernovas called Type Ia, which, due to the known luminosity of the explosion, are used as standard candles to determine distances to far off galaxies. They have been critical in the discovery of dark energy that appears to be driving the expansion rate of the universe. Consequently, astronomers are pretty attentive when a nearby white dwarf with a red giant companion is acting up, producing a nova.

RS Ophiuchi is one such system, and recently went "nova", due to a massive amount of hydrogen being dumped onto the white dwarf's surface, reaching critical mass, and undergoing a thermonuclear explosion. Scientists believe RS Ophiuchi may one day erupt in a fully-fledged supernova. If it accumulates enough mass, the white dwarf could collapse and then explode. Such explosions are triggered when all white dwarfs reach a similar mass, and therefore reach similar peak luminosity.

The surprise in observations using the Very Long Baseline Array (VLBA) shows the outburst did not appear to be symmetrical.

"We have seen structure in the blast earlier than in any other stellar explosion," said Tim O'Brien of the University of Manchester, England. "We see evidence that the explosion may be ejecting material in jets, contrary to theoretical models that assumed a spherical shell of ejected material," O'Brien added.

The nova took place inside a hot stellar wind produced by the companion red giant, and X-ray observations tracked the interaction with the expanding nova.

"We think the white dwarf in RS Ophiuchi is about as massive as a white dwarf can get, and so is close to the point when it will become a supernova," said Jennifer Sokoloski, of the Harvard-Smithsonian Center for Astrophysics. "If astronomers use such supernovas to measure the Universe, it's important to fully understand how these systems evolve prior to the explosion," she added.

The jet-like feature in the explosion leads astronomers to conclude such explosions are far

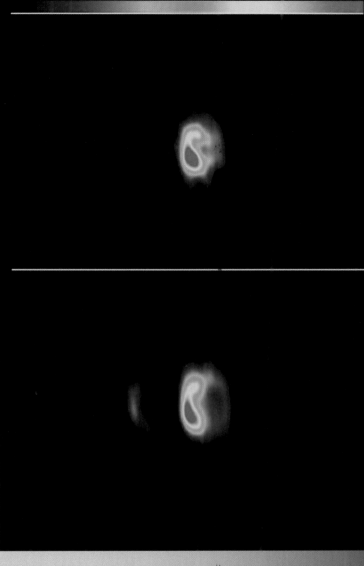

■ **Above** *19 July 2006. VLBA radio images taken 6 days apart of the white dwarf, RS Ophiuchi, caught detailed changes in the recent nova explosion. The two images show the expanding shell of debris 21 days (top) and 27 days (bottom) after the explosion. The shell spans about 30 times the Earth-Sun distance, a very tiny distance considering the 5,200 light-years distance to the object. The "jet" to the east (left) is clearly present on day 27. While such features have been seen before, the ultra-high resolution of the VLBA technique offers the highest detail yet of such outbursts. Images courtesy M. Rupen, A. Mioduszewski & J. Sokoloski, NRAO/AUI/NSF.*

more complex than previously surmised, and it's teaching new things about the physics of novas. If RS Ophiuchi truly is a Type Ia supernova waiting to happen, these detailed studies will enable better analysis of more distant Type Ia supernovae, and consequently improve our understanding of the large scale structure of the universe.

http://www.nrao.edu/pr/2006/rsoph/
http://arxiv.org/abs/astro-ph/0605326
http://arxiv.org/abs/astro-ph/0612099

21 July 2006

Giant IR Survey Revealing the Secrets of Stars and Galaxies

The largest infrared sky survey since the 2MASS (Two Micron All Sky Survey) is well underway, and scientists claim it will do for the infrared sky what the Sloan Digital Sky Survey has done for the optical sky.

The first year worth of data from a mammoth seven year project to survey the infrared sky has been released by British astronomers. Beginning in May 2005, The UKIRT Infrared Deep Sky Survey (UKIDSS) collects 150 gigabytes of data every night from the UK Infrared Telescope (UKIRT) atop Mauna Kea, Hawaii combined with the Wide Field Camera, the largest infrared imager in the world.

Initial results are now appearing from the 130 members of the survey that includes astronomers from the European Southern Observatory and some Japanese scientists.

Among several hundred thousand galaxies in the first data set, nine are massive galaxies apparently located 12 billion light-years away. Finding such fully-formed galaxies within the first billion years after the Big Bang challenges current models on how the initial phase of galaxy formation took place. Larger galaxies are thought to have formed over longer periods of time by mergers of smaller components.

"We're surveying an enormous volume of the distant Universe, which allows us to discover rare massive galaxies that were previously almost impossible to find. Understanding how these galaxies form is one of the Holy Grails of modern astronomy, and now we can trace them back to the edge of the known Universe," said survey lead, Dr. Omar Almaini of the University of Nottingham, England.

Closer to home, the survey is also detecting hundreds of brown dwarfs, offering new insights to the link between the formation of the lightest stars and the heaviest planets. Brown dwarfs are not massive enough to generate nuclear fusion in their cores, but are too large to be considered planets. As Nigel Hambly, of the Institute for Astronomy at the University of Edinburgh, hopes, "With UKIDSS, we will find many thousands of brown dwarfs in many different star formation environments within our own Galaxy."

http://www.pparc.ac.uk/Nw/ukidss2107.asp
http://www.nottingham.ac.uk/~ppzoa/UDS/
http://arxiv.org/abs/astro-ph/0610191
http://arxiv.org/abs/astro-ph/0604426

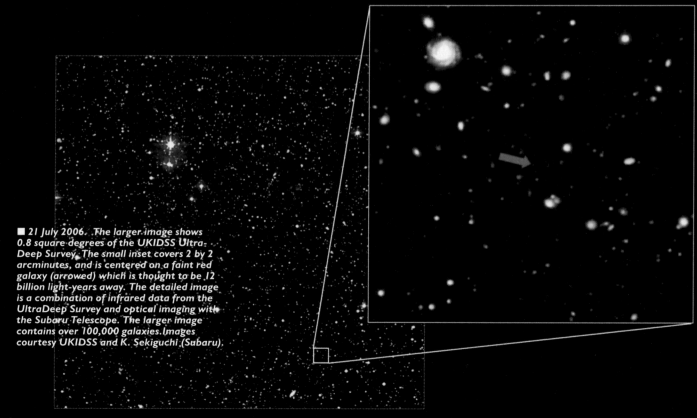

■ 21 July 2006. The larger image shows 0.8 square degrees of the UKIDSS Ultra-Deep Survey. The small inset covers 2 by 2 arcminutes, and is centered on a faint red galaxy (arrowed) which is thought to be 12 billion light-years away. The detailed image is a combination of infrared data from the UltraDeep Survey and optical imaging with the Subaru Telescope. The larger image contains over 100,000 galaxies. Images courtesy UKIDSS and K. Sekiguchi (Subaru).

25 July 2006

Surprising New Picture of Quasar Emerges

Quasars are extremely bright cores of very distant galaxies. The origin of their incredible luminosity has been long understood to be due to a central supermassive black hole. Some of the hot gas spiraling into the black hole is ejected in two opposing high velocity jets. However, studying the cores of these massive objects is difficult due to their remoteness and relatively small size.

New studies of a quasar by Rudy Schild, from the Harvard Smithsonian Center for Astrophysics, have indeed found a compact object at the center. Indeed, its mass is estimated to be 3 to 4 billion times the mass of the Sun. It has the look and feel of a black hole, but Schild has found evidence suggesting one is unlikely.

"We don't call this object a black hole because we have found evidence that it contains an internally anchored magnetic field that penetrates right through the surface of the collapsed central object, and that interacts with the quasar environment," commented Schild.

The quasar, called Q0957+561, lies behind some intervening galaxies acting as a gravitational lens. Schild and colleagues have monitored the quasar for over 20 years, watching in particular for minute changes in brightness, called microlensing events, caused by objects in the intervening galaxy pass directly between the quasar and Earth.

Such careful monitoring has allowed the astronomers, for the first time, to determine precisely where the jets form, aided by the magnifying effects of the lensing galaxy. They originate about 1,200 billion kilometers above the poles of the central compact object, a distance equivalent to about 44 light-days. However, this also implies that powering of the jets is done by energetic reconnection of tightly wound magnetic field lines that are locked to the rotating compact object.

The prevailing wisdom for quasar models is that the magnetic field is part of the accretion disk. A new model proposed by Schild and colleagues called Magnetospheric Eternally Collapsing Objects, or MECOs. The new model proposes that the quasar's central compact object may have physical properties more like a highly redshifted, spinning magnetic dipole than like a black hole.

http://www.physorg.com/news73057202.html

http://arxiv.org/abs/astro-ph/0505518

AUGUST 2006

1 August 2006

Distance to M33 Changes, and Universe Gets Bigger

The distance to one of the best known galaxies in the night sky, M33, in the constellation of Triangulum, turns out to be 15 percent farther away than previously measured. The measurement comes from a different technique to the standard Cepheid variable method.

The Cepheid method depends upon the calibration of the period luminosity law. This in turn relies on the distance to the Large Magellanic Cloud, and is fundamental in underpinning the HST Key Project distances. The new distance of 3 million light-years, compared to the previous estimate of 2.6 million light-years, was determined using a rare type of eclipsing binary star in M33.

The measurement was conducted in a 10 year-long project called DIRECT, led by Kris Stanek, associate professor of astronomy at Ohio State University, and Alceste Bonanos of the Carnegie Institute of Washington. The goal of the DIRECT project is to use detached eclipsing binaries to determine the distances of M31 and M33 to better than 5 percent accuracy compared to the present 15 percent accuracy.

Finding two stars in a binary system enables astronomers to determine the mass of each star. Once that is known, their actual luminosity can be calculated, and compared with the apparent brightness as seen from Earth, which leads to a direct distance determination.

Technical challenges in observing stars in M33, which are six magnitudes fainter than LMC stars, led to the long development time for the new method. Spectroscopic measurements critical to this observation are more difficult, with fainter stars and a more crowded star field to deal with in galaxies at greater distances than the LMC. A wide range of telescopes from 1-meter class to the Keck 10-meter were used for the observations.

If this new distance measurement is correct, then the true value of the Hubble constant may be 15 percent smaller – and the universe may be 15 percent bigger and older – than previously thought. Such a radical change requires many more observations to be undertaken. There are three more eclipsing binaries under scrutiny in M33, hopefully helping astronomers to understand why the Cepheid and eclipsing binary distances are so different.

http://cfa-www.harvard.edu/~kstanek/DIRECT/

http://arxiv.org/abs/astro-ph/0606279

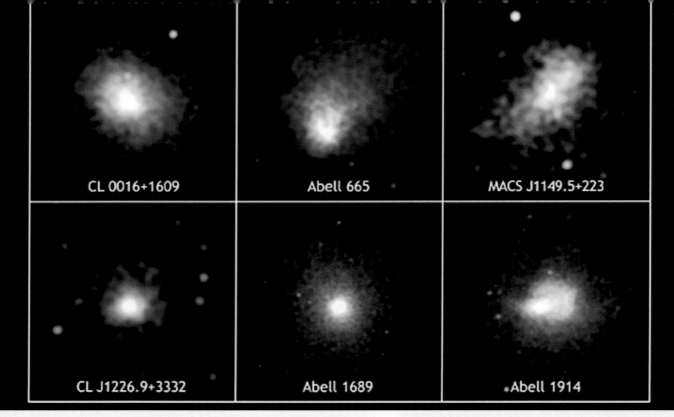

CL 0016+1609   Abell 665   MACS J1149.5+223

CL J1226.9+3332   Abell 1689   Abell 1914

8 August 2006

Chandra Independently Determines Hubble Constant

Knowing the rate at which the universe is expanding is arguably the most important number in cosmology. The rate, called the Hubble constant, is known to within 15 percent accuracy. New methods independent of the traditional Cepheid period-luminosity law are being tested, and one such method has produced some valuable results, independently confirming results from the Hubble Space Telescope.

The method uses a combination of radio and X-ray observations of clusters of galaxies. Astronomers have known since the early 1970s that photons from the cosmic microwave background (CMB) are affected by hot gas that pervades giant galaxy clusters. This phenomenon, known as the Sunyaev-Zeldovitch effect, produces a tiny distortion of about 1 milli-kelvin on CMB photons. The magnitude depends on the density and temperature of the hot electrons in the gas and the physical size of the cluster.

The Owens Valley Radio Observatory and the BIMA array (now combined in the CARMA array) radio telescopes measured the S-Z effect, and data from the Chandra X-ray Observatory measured the properties of the hot gas. When combined, the physical size of the cluster may be calculated. Using the angular size of the cluster on the sky, a simple calculation gives the distance to the cluster, independent of any standard candles. The Hubble constant comes from dividing this distance by the recession velocity of the cluster.

The results from 38 galaxy clusters give a Hubble constant of 77 kilometers per second per megaparsec, with an uncertainty of 15 percent. The galaxy clusters range from 1.4 billion to 9.3 billion light-years.

The new result confirms previous measurements and fixes the age of the universe at between 12 and 14 billion years. The importance of the new result lies in the fact that an entirely different set of assumptions to the traditional standard candle method are used. "These new results are entirely independent of all previous methods of measuring the Hubble constant," said Dr. Marshall Joy of NASA's Marshall Space Flight Center.

http://chandra.harvard.edu/press/06_releases/press_080806.html

http://arxiv.org/abs/astro-ph/0512349

■ *Above: 8 August 2006. Six of the thirty eight galaxy clusters studied by the Chandra X-ray Observatory are shown here, with distances ranging from 1.4 to 9.3 billion light years from Earth. The observations were part of a campaign to determine the Hubble constant using the so-called Sunyaev-Zeldovich effect. Chandra measured the hot intergalactic gas which is used to determine the physical size, and therefore distance, to each cluster.*
Image courtesy NASA/CXC/MSFC/M.Bonamente et al.

14 August 2006

NASA's FUSE Satellite Deciphers Key Tracer of Galaxy Evolution

The Far Ultraviolet Spectroscopic Explorer (FUSE) has determined the deuterium (heavy hydrogen) content in our Milky Way to a higher precision than any previous measurement. To their surprise, scientists found more of it than expected.

The amount of deuterium plays a critical role in the early history of the universe. Except for a brief post-Big Bang time period, there's no known mechanism for creating deuterium. It is easily destroyed inside stars, leading scientists to conclude that deuterium is primordial in origin, and therefore should be present in the same levels everywhere.

Ever since the 1970s, however, deuterium levels have been found to be quite variable, presenting a puzzle for theorists. Now the puzzle may be solved by the 2003 idea of Bruce Blaine from Princeton University. He explained that deuterium might preferentially bind to interstellar dust grains in favor of hydrogen, and thereby changing into an undetectable solid form. If so, there would be higher levels of deuterium found in regions of supernova explosions, or hot stellar winds, since the warming would release the molecule from dust grains.

Indeed, this is just what FUSE finds, giving support for Blaine's model. However, while it does appear to explain the varying levels, scientists remain puzzled over the average level of deuterium remaining. Primordial levels are around 27 parts per million, and over 30 percent is expected to have been destroyed due to star formation. FUSE indicates only 15 percent has been destroyed, suggesting less than expected has been cycled through stars. The chemical evolution of our Milky Way requires revision to explain this result.

"FUSE has solved the mystery about why the deuterium is where it is, but now scientists need to try to explain why there is so much of it," said Brian Fields of the University of Illinois in Urbana. Team leader Jeffery Linsky at JILA, University of Colorado, sums it up by adding,

"Since the 1970s we have been unable to explain why deuterium levels vary all over the place," he said. "The answer we have found is as unsettling as it is exciting."

http://fuse.pha.jhu.edu/wpb/science.html
http://arxiv.org/abs/astro-ph/0608308

15 August 2006

Spitzer Space Telescope Discovers New Wonders in the Orion Nebula

A dramatic infrared view of the famous Orion Nebula by the Spitzer Space Telescope (SST) has revealed swirls of gas and hundreds of previously unseen stars, some with planetary disks. The new view, made by combining about 10,000 individual exposures, complements last year's release of a large mosaic taken by the Hubble Space Telescope in optical wavelengths.

The survey of the Orion star forming region is part of detailed survey of young star-forming

■ **Above:** *15 August 2006. A comparison of the infrared (left) and optical (right) views of the Orion Nebula reveals the power of viewing longer wavelengths. Infrared penetrates the thick columns of dust that otherwise obscure optical wavelengths. Nestled inside these dark cocoons are new stars forming. The Spitzer Space Telescope took a series of images using the Infrared Array Camera to create the montage. Light with wavelengths of 8 and 5.8 microns (red and orange) comes mainly from dust that has been heated by starlight. Light of 4.5 microns (green) shows hot gas and dust; and light of 3.6 microns (blue) is from starlight. Image courtesy NASA/JPL-Caltech/ T. Megeath (University of Toledo) – Infrared; NOAO/AURA/NSF/A. Block/R. Steinberg – Visible.*

regions being led by Tom Megeath of the University of Toledo. "I was immediately struck by the intricate structure in the nebulosity," he said. The Orion Nebula is the nearest star forming region to Earth, located about 1,450 light-years away. Infrared observations allow astronomers to see through some of the dense clouds to see what processes are going on deep within the nebula. Spitzer found 2,300 planet-forming disks through their tell-tale infrared excess in the spectrum produced by dust. Most are too small to be seen in an image like this.

"Most stars form in crowded environments like Orion, so if we want to understand how stars form, we need to understand the Orion nebula star cluster," explained Dr. Lori Allen of the Center for Astrophysics at Harvard University.

http://arxiv.org/abs/astro-ph/0701476

http://www.spitzer.caltech.edu/Media/ releases/ssc2006-16/index.shtml

■ *Below: 16 August 2006. SINFONI is a very powerful instrument being used on the European Southern Observatory's Very Large Telescope. It can acquire spectra of 2,000 points within a single field of view simultaneously, allowing previously unattainable detail of motions within distant galaxies to be determined. Part of the SINFONI installation is shown in its early commissioning phase attached to the 8.2-meter VLT Yepun. SINFONI comes in two parts, a high resolution infrared spectrograph, and an adaptive optics module. The blue ring is the Adaptive Optics Module. The yellow parts simulate the spectroscopic component.*
Image courtesy European Southern Observatory.

16 August 2006

SINFONI Discovers Rapidly Forming, Large Proto-Disk Galaxies Three Billion Years After the Big Bang

A combination of adaptive optics and a high resolution spectrograph on the European Southern Observatory's 8.2-meter Very Large Telescope has been able to extract stunning details in a very distant proto-disk galaxy. The galaxy appears similar to the Milky Way, and must have formed only 3 billion years after the Big Bang.

The challenging task before astronomers is to explain how large galaxies formed so soon after the Big Bang. To answer the question, high resolution data from distant small specks of light are required, demanding the best out of the world's largest telescopes. SINFONI is an infrared spectrometer combined with an adaptive optics module that can meet the challenge. It can collect spectra from 2,000 points simultaneously in a single field of view.

By using the SINFONI instrument on the VLT, astronomers achieved an angular resolution of 0.15", corresponding to 4,000 light-years at the distance of the galaxy. The galaxy in question, called BzK155043, lies at a redshift of $z = 2.4$.

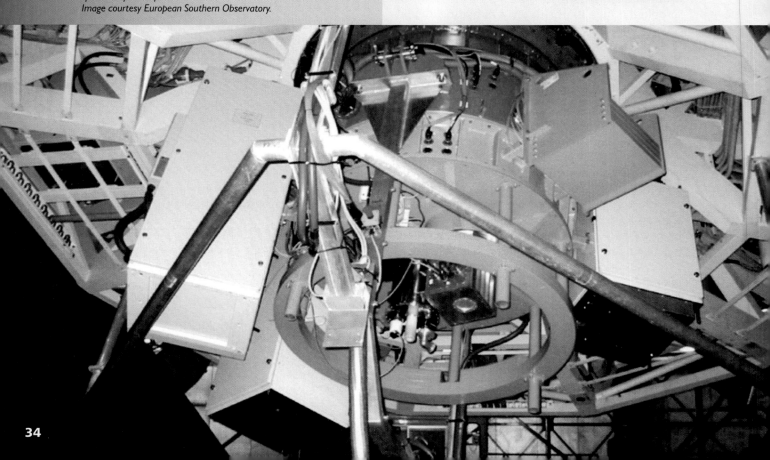

■ *21 August 2006. The so-called "Bullet cluster" is a composite of X-ray, optical and inferred mass from gravitational lensing, providing strong evidence for the presence of dark matter. The optical image from the Magellan and the Hubble Space Telescope shows galaxies in orange and white. Hot gas, which contains the bulk of the normal matter in the cluster, is shown by the Chandra X-ray image (pink). Gravitational lensing, the distortion of background images by the total mass (ordinary plus dark matter) in the cluster, is shown in blue. Because dark matter doesn't interact very much with ordinary matter, the blue distribution is separate from the ordinary matter distribution, giving the strongest evidence yet that nearly all of the matter in the clusters is dark. Image courtesy NASA/CXC/CfA/M.Markevitch et al. – X-ray; NASA/STScI and Magellan/U.Arizona/ D.Clowe et al. – Optical; NASA/STScI, ESO WFI and Magellan/U.Arizona/D.Clowe et al. – Lensing Map.*

"We have been able, for the first time, to obtain well resolved, two-dimensional images of the gas motions in distant star forming galaxies, whose light has traveled more than 11 billion years to the Earth," said Reinhard Genzel of the Max-Planck-Institute for Extraterrestrial Physics, Garching, Germany.

The observations are part of a large project called SINS (Spectroscopic Imaging Survey in the Near-Infrared with SINFONI), designed to study distant, luminous star-forming galaxies with the VLT. "When we started the SINS program we expected to see mostly irregular and perhaps even chaotic motions caused by the frequent merger activity in the young universe," Genzel added. "We were in for a major surprise when we found a number of large, rotating, and gas rich disk galaxies whose properties are quite similar to the present day Milky Way," he said.

"We have a growing body of evidence that massive galaxies formed much more rapidly in the redshift range 2-3 than originally anticipated," said Andrea Cimatti of the University of Bologna. "The new SINFONI data give us a first glimpse what processes might be involved."

http://www.eso.org/outreach/press-rel/pr-2006/pr-31-06.html

http://www.eso.org/outreach/press-rel/pr-2004/pr-21-04.html

http://arxiv.org/abs/astro-ph/0608344

21 August 2006

Chandra X-Ray Observatory Finds Strong Evidence of Dark Matter

A wide range of telescopes covering optical and X-ray wavelengths have observed a collision between a small galaxy cluster and a larger one, and deduced that dark matter must be present in order to explain the observations. The Hubble Space Telescope, the European Southern Observatory's Very Large Telescope, and the Magellan telescopes joined the orbiting Chandra X-ray Observatory for the campaign.

The technique used was first to determine the precise location and distribution of mass in the cluster by observing how light from more distant galaxies is distorted by gravitational lensing. This reveals the total mass of the cluster, the sum of ordinary and dark matter. The next step is to measure the location and distribution of ordinary matter from X-ray observations. X-rays highlight the hot gas from the collision and the hot intergalactic gas that pervades both clusters.

Because the two clusters are moving during the collision, the dark matter, which doesn't interact with ordinary matter, will have moved father than ordinary matter, and so the locations of the measured mass by the two methods will show a difference.

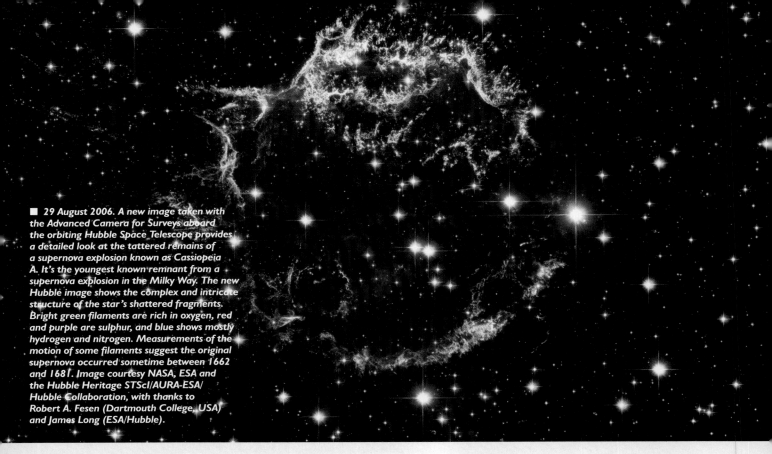

■ 29 August 2006. A new image taken with the Advanced Camera for Surveys aboard the orbiting Hubble Space Telescope provides a detailed look at the tattered remains of a supernova explosion known as Cassiopeia A. It's the youngest known remnant from a supernova explosion in the Milky Way. The new Hubble image shows the complex and intricate structure of the star's shattered fragments. Bright green filaments are rich in oxygen, red and purple are sulphur, and blue shows mostly hydrogen and nitrogen. Measurements of the motion of some filaments suggest the original supernova occurred sometime between 1662 and 1681. Image courtesy NASA, ESA and the Hubble Heritage STScI/AURA-ESA/ Hubble Collaboration, with thanks to Robert A. Fesen (Dartmouth College, USA) and James Long (ESA/Hubble).

And that's precisely what the observations found. The galaxy cluster 1E0657-56, also called the "Bullet cluster", because it contains a spectacular bullet-shaped cloud of hundred-million-degree gas, was the target of 100 hours of observation by Chandra. The bullet shape is produced because of the high-speed collision of the hot gas in each cluster.

The hot gas in this collision was slowed by a drag force, similar to air resistance. In contrast, the dark matter was not slowed by the impact, because it does not interact directly with itself or the gas, except through gravity. The result is the observed separation of the deduced dark matter distribution and normal matter seen in the image.

"A universe that's dominated by dark stuff seems preposterous, so we wanted to test whether there were any basic flaws in our thinking," said Doug Clowe of the University of Arizona at Tucson, and leader of the study. The results are the best evidence to date in support of dark matter.

Alternative theories of gravity that modify the present laws of gravity to explain observations, predict that such an observed separation between the gravitational-lens-induced mass and the hot gas would not occur.

"This is the type of result that future theories will have to take into account," said Sean Carroll, an independent cosmologist at the University of

Chicago. "As we move forward to understand the true nature of dark matter, this new result will be impossible to ignore."

http://arxiv.org/abs/astro-ph/0611496
http://chandra.harvard.edu/photo/2006/1e0657/

29 August 2006

Cassiopeia A - The Colorful Aftermath of a Violent Stellar Death

Cassiopeia A is a well-known supernova remnant, the youngest known in the Milky Way. A new image from the Hubble Space Telescope of Cassiopeia A reveals swirls of gas, clumps, and filaments, strung out by the rush of the passing shockwave from the original supernova explosion. It lies 10,000 light-years away.

HST, like other telescopes, produces color images by adding many black and white images obtained through selected color filters that isolate certain chemical species. Each image is later stacked in layers, with color added to each layer representing the various chemical elements. In the HST image, taken by the Advanced Camera for Surveys, oxygen is depicted as bright green, sulphur is shown as red and purple, while hydrogen and nitrogen are both blue.

Hubble has seen changes in the nebula in images taken nine months apart. Along the upper left side of the expanding nebula is faint debris rushing along at 50 million kilometers per hour (31 million mph).

Studies of the motion of 1,800 knots, by Robert Fesen of Dartmouth College and colleagues, suggest a date for the initial explosion as occurring in 1681, although some knots show slightly different motion, suggesting it could have been as early as 1662. Earlier estimates had arrived at a 1671 date. The disparity comes from the unknown deceleration of knots, though future imaging by ACS and other telescopes may narrow down the date further.

http://arxiv.org/abs/astro-ph/0603371

http://arxiv.org/abs/astro-ph/0509067

http://arxiv.org/abs/astro-ph/0509552

http://www.spacetelescope.org/news/html/heic0609.html

31 August 2006

Spitzer Maps Evolution in the Large Magellanic Cloud

In a massive effort, 600,000 images from the Spitzer Space Telescope's Multiband Imaging Photometer (MIPS) have been combined to produce a magnificent mosaic of the Large Magellanic Cloud (LMC). The image is part of a huge effort by Spitzer astronomers called "Surveying the Agents of a Galaxy's Evolution" (SAGE).

The LMC is an excellent target for such a study, because the galaxy has a favorable viewing angle, and most clouds are seen as single clouds, whereas in the Milky Way the line of site is confused by many infrared sources. In addition, all the objects in the LMC can be considered to be the same distance away, simplifying direct comparisons between different regions in the galaxy.

"We can use this amazing map to really start to understand in detail how a galaxy evolves," said Karl Gordon of the University of Arizona Steward Observatory. "We can now test sophisticated theories about how stars form, how they evolve, what the different populations

are, and how important they are in a global galaxy environment," Gordon added.

Dr. Margaret Meixner of the Space Telescope Science Institute is another member of the large group of astronomers involved in the effort to understand how stardust is recycled in galaxies like the LMC. "The Large Magellanic Cloud is like an open book," said Meixner. "We can see the entire lifecycle of matter in a galaxy in this one snapshot."

http://arxiv.org/abs/astro-ph/0606356
http://arxiv.org/abs/astro-ph/0608189
http://sage.stsci.edu/
http://www.spitzer.caltech.edu/Media/releases/ssc2006-17/

■ **Below:** *31 August 2006. Ten thousand individual images from the Spitzer Space Telescope were used to create this magnificent mosaic of our satellite galaxy, the Large Magellanic Cloud (LMC). With its infrared detectors piercing through some of the dark clouds present in the LMC, Spitzer has identified nearly one million new objects. Blue represents light from older stars. The red clouds show interstellar gas and dust illuminated by starlight. This picture is a composite of infrared light captured by Spitzer's infrared array camera. Light with wavelengths of 8 and 5.8 microns is red and orange; 4.5-micron light is green; and 3.6-micron light is blue. Image courtesy NASA/JPL-Caltech/M. Meixner (STScI) & the SAGE Legacy Team.*

SEPTEMBER 2006

8 September 2006

Jupiter-Sized Transiting Planet Found by Astronomers Using Novel Telescope Network

Telescopes similar to those found in the hands of many amateur astronomers have discovered a planet transiting a star located 500 light-years from Earth. It's the second success for the Trans-Atlantic Exoplanet Survey (TrES). The transiting object, called TrES-2, crosses in front of its parent star every two-and-a-half days. The TrES uses several 10-cm telescopes at different locations designed for automated sequences of observations.

The discovery is particularly noteworthy because it lies within the planned field of view of the forthcoming Kepler mission, a spacecraft designed to search for transiting planets. Kepler's mission will be able to observe about 600 transits of TrES-2 during its mission lifetime of three years.

TrES-2 weighs in at about 1.3 times the mass of Jupiter and orbits a yellow star about the mass of the Sun. Francis O'Donovan, a graduate student in astronomy at the California Institute of Technology, explains how the telescopes spotted the new planet. "When TrES-2 is in front of the star, it blocks off about one and a half percent of the star's light, an effect we can observe with our TrES telescopes," he said. The new extrasolar planet was confirmed by follow-up observations with the 10-meter Keck telescope.

http://www.lowell.edu/press_room/releases/recent_releases/TrES2.html
http://arxiv.org/abs/astro-ph/0609335

13 September 2006

Astronomers Trace the Evolution of the First Galaxies in the Universe

Astronomers trying to find the earliest galaxies that formed after the Big Bang have found an interesting peak in their frequency of formation. About 900 million years after the Big Bang, hundreds of bright galaxies were visible. Yet 200 million years earlier there seem to be very few. As one of the astronomers, Garth Illingworth at the University of California, Santa Cruz, commented, "There must have been a lot of merging of smaller galaxies during that time."

The observations come from a study of the deepest images ever taken by the Advanced Camera for Surveys aboard the Hubble Space Telescope. Scanning images obtained for the Hubble Ultra-Deep Field and the Great Observatories Origins Deep Survey (GOODS) fields, astronomers searched for galaxies near the limit of visibility. These objects would likely be the farthest away, and consequently, due to

■ *8 September 2006. A computer-generated simulation of the transiting planet, TrES-2, shown crossing in front of the disk of its parent star. The transit causes a drop in the brightness of about one and a half percent, and was detected during automated observations from a 10-cm telescope of the Trans-Atlantic Exoplanet Survey. The new planet is estimated at 1.3 times the mass of Jupiter. Image courtesy Jeffery Hall, Lowell Observatory.*

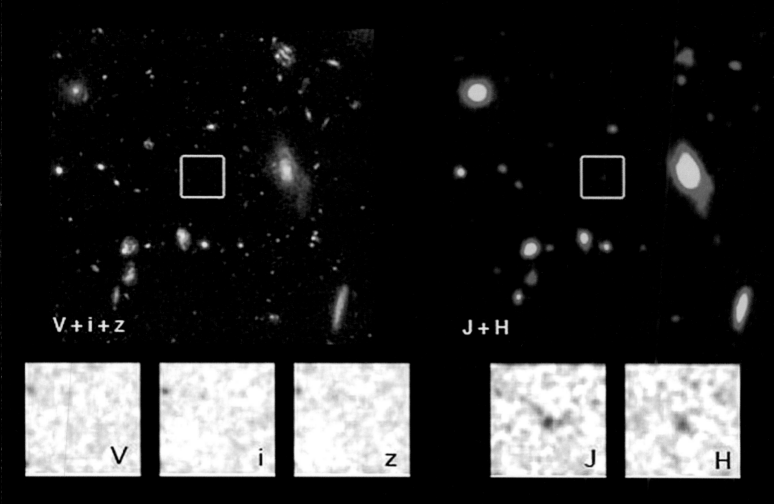

V + i + z

J + H

V

i

z

J

H

the finite speed of light, be some of the earliest galaxies to have formed following the Big Bang 13.7 million years ago.

Astronomers used the observed sample of galaxies seen 900 million years after the Big Bang, and ran a simulation of how many should be visible 200 million years earlier, or 200 million light-years farther away. The prediction, assuming the galaxies were of the same type at this earlier epoch, was that between 10 and 17 sources should be visible in the area of sky under observation. However, only four galaxies are visible. This suggests such luminous galaxies were quite rare 700 million years after the Big Bang, and a period of rapid growth must have occurred to account for the observed number of galaxies in the original sample. The four galaxies appear much smaller than our Milky Way galaxy, supporting the theory of hierarchical growth of galaxies. It's as if many mergers of small galaxies occurred during this 200 million year period of the early universe.

http://firstgalaxies.ucolick.org/

http://arxiv.org/abs/astro-ph/0607087

13 September 2006
Cosmic Archeology Uncovers the Universe's Dark Ages

The 8-meter Subaru telescope is at the forefront of finding the most distant galaxies ever seen. One of its key advantages is the ability to locate its wide-field camera at the prime focus of the giant telescope, the only 8-meter class telescope with this ability.

Astronomers from the National Astronomical Observatory of Japan (NAOJ) announced their most recent find using Subaru, a galaxy called IOK-1. Its spectrum shows it to be at a redshift of 6.96, placing the galaxy 12.88 billion light-years from Earth. This discovery and others like it suggest these galaxies existed only 780 million years after the Big Bang.

■ *Above: 13 September 2006. Images of a galaxy at redshift 7.4 (inside white box) in the Hubble Ultra Deep Field. This galaxy is seen just 700 million years after the Big Bang. The galaxy disappears at optical wavelengths (left), but is seen clearly in the infrared (right), as shown in the image boxes at the bottom (shown in infrared J and K wavelength bands.) Image courtesy Bouwens/Magee.*

However, like the University of California study in the previous story, based on the number of galaxies observed at the later epoch of 840 million years after the Big Bang, as many as six galaxies should have been visible in the field of view. I0K-1 is only one of two candidates for galaxies at the earlier epoch. It appears the universe changed during the 60 million year gap.

Masanori Iye and Nobunari Kashikawa (NAOJ) and Kazuaki Ota of the University of Tokyo, suggest the most exciting conclusion is that we are seeing the epoch of reionization. Clouds of neutral hydrogen blocking our view of the earlier universe became transparent due to the ionizing effects of a large number of stars. 780 million years after the Big Bang, there were not enough stars to ionize all the hydrogen, but 60 million years later that situation changed.

This exciting conclusion requires further work and complements the idea that rapid hierarchical growth occurred. The larger infrared telescopes planned for the next decade will likely provide more definitive answers.

http://www.naoj.org/Pressrelease/2006/09/13/index.html
http://arxiv.org/abs/astro-ph/0609393

20 September 2006
"Champagne Supernova" Challenges Understanding of How Supernovas Work

When observations challenge some of our basic understanding of how the universe works, many scientists see it as a cause for celebration, hence the label "champagne". Our understanding of Type Ia supernovas, in which a white dwarf explodes, is based upon the long standing knowledge of the Chandrasekhar limit. The limit determines the critical mass above which a white dwarf will explode. Based on solid physical laws the value is 1.4 times the mass of the Sun. When a white dwarf is a member of a binary system, and its companion is a red giant star, the white dwarf can draw material from the red giant, eventually tipping over the Chandrasekhar limit, and exploding as a Type Ia supernova.

When a Type Ia supernova like SNLS-03D3bb turns up in a distant galaxy four billion light-years away, it's left to astronomers like Andy Howell at the University of Toronto to determine the mass of the original star. However, this one, discovered by the SuperNova Legacy Survey, gave a surprising result. The white dwarf was more massive than any previous example - fifty percent more. Such a result is disturbing, because Type Ia events are used as reliable distance indicators because the explosion is thought to involve a standard amount of fuel.

The question is, how could a white dwarf accumulate mass above the Chandrasekhar Limit without exploding? A number of scenarios have been suggested. Perhaps it was a rare case of two white dwarfs colliding. Alternatively, the carbon and oxygen ratio in the white dwarf may be different than normal, producing a greater quantity of nickel in the explosion, causing a much brighter supernova. However, a larger, brighter, explosion would generate higher velocity debris, and the SNLS-03D3bb debris was slower. Another alternative is the progenitor was spinning rapidly, counterbalancing the gravitational forces and allowing the white dwarf to sustain greater mass before exploding.

There is already an indication of two population types leading to Type Ia supernovas. As colleague Peter Nugent of the Lawrence Berkeley National Laboratory explains, "They can crudely be broken into those that come from young star-forming galaxies and those from old, dead galaxies".

Young star-forming galaxies produce massive objects and could be rich in white-dwarf plus white-dwarf binary systems, so-called "double-degenerate" systems.

"If the double-degenerate model is right, such systems will always produce super-Chandrasekhar explosions in these very young galaxies," Nugent added. Young galaxies are found at greater distances, and Type Ia supernovas at large distances are used as evidence for dark energy. Understanding the different types of white dwarf explosions is essential to cosmology.

http://www.lbl.gov/Science-Articles/Archive/Phys-weird-supernova.html
http://www.cfht.hawaii.edu/SNLS/
http://arxiv.org/abs/astro-ph/0609616

27 September 2006

Astronomers Gain Important Insight into How Massive Stars Form

Modeling the formation of massive stars has repeatedly hit a snag. Stars close to 8 times the mass of the Sun generate so much radiation during their formation that it prevents further infall of gas. As Maria Teresa Beltran of the University of Barcelona in Spain explains, "We know there are many stars bigger than that, so the question is, how do they get that much mass?"

New observations by the Very Large Array in Socorro, New Mexico, has discovered new evidence supporting an idea that infalling matter forms a disk, and most of the radiation escapes without hitting the disk.

Beltran and other astronomers from Italy and Hawaii studied a young, massive star called G24 A1 that tips the scale at 20 times the mass of the Sun. It's located about 25,000 light-years from Earth. If the disk model is true, observations should show material falling in, rushing out, and orbiting, all at the same time.

Astronomers used the VLA to observe radio emissions from the ammonia molecule around G24 A1 near a frequency of 23 GHz. Doppler shifts in the observed frequency reveal internal motions, allowing astronomers to build a model of what is going on around the young massive star. Beltran announced, "It's the first time all three types of motion have been seen in a single young massive star. Our findings suggest that the disk model is a plausible way to make stars up to 20 times the mass of the Sun," she said.

http://www.nrao.edu/pr/2006/starflow/
http://arxiv.org/abs/astro-ph/0609789

28 September 2006

Anatomy of a Planet-Forming Disk Around a Star More Massive than the Sun

Planet-forming debris disks around stars are numerous, yet astronomers would like to know more about what is going on inside the disk. Using advanced instrumentation on the European Southern Observatory's Very Large Telescope (VLT), such a goal is now achievable. The instrument is called VISIR, the VLT Imager and Spectrometer for the InfraRed. It can sample images down to the diffraction limit of the 8.2-meter Melipal telescope (0.3 arcseconds at 10 microns wavelength).

Pierre-Olivier Lagage of the French Atomic Energy Commission in Saclay and colleagues used VISIR to map the disk of the young star HD 97048. It lies in the Chameleon I dark cloud located 600 light-years from Earth. At the angular resolution of VISIR, astronomers were able to resolve a very large disk spanning 12 times farther from the star than Neptune is from our Sun, and remarkably also noticed that the outer edge of the disk appears to be flared. "This is the first time such a structure, predicted

■ 27 September 2006. Artist's impression of a young star showing the multiple motions detected in a star called G24 A1. Infalling matter forms a disk, which is rotating, and the outflow of hot gas and radiation misses the main disk, allowing massive stars greater than 8 solar masses to form. Image courtesy Bill Saxton, NRAO/AUI/NSF.

by some theoretical models, is imaged around a massive star," said Lagage.

The observations lead to an estimate of at least 10 times the mass of Jupiter in the form of gas and more than 50 Earth masses in dust. The dust level is a thousand times more than found around stars like Fomalhaut and Vega, which are older stars. Given that dust around older stars may already have formed planets, what astronomers are seeing around HD 97048 is likely a dust disk at an earlier stage of evolution.

"From the structure of the disk, we infer that planetary embryos may be present in the inner part of the disk," said Lagage. "We are planning follow-up observations at higher angular resolution with ESO's VLT interferometer in order to probe these regions."

http://arxiv.org/abs/astro-ph/0610322
http://www.eso.org/outreach/press-rel/pr-2006/pr-36-06.html

OCTOBER 2006

3 October 2006

Astronomers Construct Largest 3D All-Sky Map of Galaxies and Motions

With a map of 25,000 galaxies out to 600 million light-years, astronomers have completed the most detailed map of the entire sky in three dimensions. The all-sky map includes all the major superclusters of galaxies and voids within this region of space. The most massive feature is the 20-million light-year long Shapley supercluster, located 400 million light-years from our galaxy. Also visible on the map is the "Great Attractor", one-third of the distance of the Shapley cluster, and it has a significant effect on the motion of our Milky Way.

Large scale galaxy surveys provide valuable information about dark matter, the elusive material than doesn't interact with normal matter, yet leaves a distinctive gravitational imprint on the universe. "We need to map the distribution of dark matter rather than luminous matter in order to understand large-scale motions in our Universe," explained Dr. Pirin Erdogdu of Nottingham University. "Fortunately on large scales, dark matter is distributed almost the same way as luminous matter, so we can use one to help unravel the other."

The new map comes from the 2MASS Redshift Survey (2MRS) and involves an international team of astronomers. The Two Micron All Sky Survey (2MASS) was a four year project to scan the entire sky at infrared wavelengths, completed in 2001. The 3D map was constructed by obtaining spectra of 25,000 galaxies. Their redshifts produce a distance for each galaxy. Working out the dark matter distribution was aided by Ofer Lahav at University College, London, who used the relationship of galaxy velocities and the total distribution of mass.

"Our nearly two decades of effort has produced the absolute best ever map of the

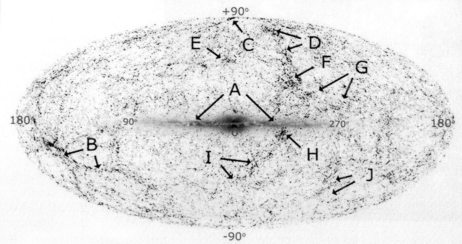

■ *Left: 3 October 2006. This panoramic view of the entire near-infrared sky reveals the distribution of galaxies beyond the Milky Way. The image is derived from the 2MASS Extended Source Catalog (XSC) — more than 1.5 million galaxies, and the Point Source Catalog (PSC) — nearly 0.5 billion Milky Way stars. The galaxies are color coded by "redshift". Blue are the nearest sources (z < 0.01); green are at moderate distances (0.01 < z < 0.04) and red are the most distant sources that 2MASS resolves (0.04 < z < 0.1). A locator key for various galaxy superclusters is shown below the panoramic view. Images courtesy Thomas Jarrett (IPAC/Caltech).*

A: Milky Way
B: Perseus-Pisces Supercluster
C: Coma Cluster
D: Virgo Cluster/Local Supercluster
E: Hercules Supercluster
F: Shapley Concentration/Abell 3558
G: Hydra-Centaurus Supercluster
H: "Great Attractor"/Abell 3627
I: Pavo-Indus Supercluster
J: Horologium-Reticulum Supercluster

nearby Universe," said John Huchra of Harvard University. "With this we hope to elucidate the nature and disposition of dark matter and understand much, much more about our cosmological model and about galaxies themselves."
http://cfa-www.harvard.edu/~huchra/2mass/
http://arxiv.org/abs/astro-ph/0610005

4 October 2006

Hubble Finds Extrasolar Planet Candidates Far Across Our Galaxy

It's becoming a regular story now. A new extrasolar planet is discovered orbiting some nearby star. There are well over 200 known. The Hubble Space Telescope, however, appears to have found sixteen, and they are not nearby. They're located halfway across the galaxy, 26,000 light-years away. They are called candidates, because only two have been confirmed by follow-up measurements.

The sixteen new candidates were found as part of the Sagittarius Window Eclipsing Extrasolar Planet Search (SWEEPS) program. Using the power of the Advanced Camera for Surveys, Hubble monitored the brightness of over 180,000 stars in the richly packed star field towards the center of the Milky Way, a region in the constellation of Sagittarius. If a planet wanders in front of a star, a so-called transit event, it causes a momentary dip in a star's brightness. It's estimated that out of every 10,000 stars surveyed, one will produce a transiting planet that is observable.

Hubble saw up to fifteen consecutive transits of each star to confirm their existence. Follow up observations for the two brightest stars using the Very Large Telescope at the European Southern Observatory confirmed a slight spectroscopic wobble in both stars, validating the detection of planets using the transit technique and enabling their masses to be estimated. One is below 3.8 Jupiter masses and the other weighs in at 9.7 Jupiter masses. Both candidates are below the minimum mass of 13 Jupiter masses for a brown dwarf. Such confirmation adds credence to the remaining 14 candidates.

Of the sixteen, five are quite unusual, orbiting their parent star extremely closely. "Discovering the very short-period planets was a big surprise,"

waiting for the next burst scanning the sky for the highest energy X-ray sources. Swift spent nine months coming up with a census of galaxies with active galactic nuclei (AGNs) that contain supermassive black holes.

The all-sky survey found more than 200 AGNs, some previously undetected due to obscuring dust and gas. Large, dense accretion disks occur in AGNs that are also starburst galaxies. The visibility of the central bright region around the black hole depends on the angle of the disk to our line of sight and some properties are hidden from view. Swift is able to detect these objects for the first time, due to the penetrating power of the highest energy X-rays.

"We are confident that we are seeing every active, supermassive black hole within 400 million light-years of Earth," said Jack Tueller of NASA's Goddard Space Flight Center in Greenbelt, Maryland. This volume of space covers two-thirds of the 2MASS redshift survey. (See also 3 October story.)

The dust enshrouded AGNs provide a test of how star formation and black holes feed each other. As Richard Mushotzky of GSFC points out, "You can't understand the universe without understanding black holes. Perhaps as much as 20 percent of all of the radiated energy in the universe arises in one way or another from AGN activity."

Swift continues to scan the sky, and each scan is added to the previous one, increasing the sensitivity of the survey. More black holes from more distant reaches of the universe will eventually be detected.

http://www.nasa.gov/centers/goddard/news/topstory/2006/blackhole_headcount.html

http://arxiv.org/abs/astro-ph/0509860

http://arxiv.org/abs/astro-ph/0612497

said team leader Kailash Sahu of the Space Telescope Science Institute, Baltimore. The so-called Ultra-Short-Period-Planets orbit their stars in less than a day.

Scientists involved with the Kepler mission, due for launch in late 2008, are excited to learn this technique works. The Kepler mission is designed to survey a similar region of the Milky Way in the search for Earth-like planets.

http://hubblesite.org/news/2006/34

http://arxiv.org/abs/astro-ph/0610098

5 October 2006

Black Holes I:
Swift Observatory Performs Headcount of Local Black Holes

In addition to detecting gamma-ray bursts, the Swift orbiting observatory spends the time

■ **Above:** *4 October 2006. This Hubble Space Telescope image shows half of the region studied that produced the dramatic discovery of 16 extrasolar planet candidates. The full field contains over 180,000 stars, and one in 10,000 are predicted to have planets that are suitably oriented to produce occasional transits, resulting in minute drops in brightness of the parent stars. The image was taken as part of the Sagittarius Window Eclipsing Extrasolar Planet Survey (SWEEPS). The bottom series shows one of two stars in the field where spectroscopic evidence revealed the tell-tale signature of a wobble in the star induced by the presence of a planet with about 3.8 times the mass of Jupiter. Image courtesy NASA, ESA, K. Sahu (STScI) and the SWEEPS Science Team.*

5 October 2006

Black Holes II: Scientists Nudge Closer to Edge of Black Hole

Discovering activity close to the edge of a black hole is tricky. The active region is so small as seen from Earth that astronomers have to look for circumstantial evidence. In the high energy, high density regime of a black hole, the X-ray region of the spectrum is the place to look. In particular, a spectral line due to iron, called the "broad iron K line", is the key to unlocking some of the mysteries around a black hole.

"Across the board, we are finding the broad iron K line to be an incredibly robust measure of black hole properties," said Andrew Fabian of Cambridge University, England. "We are entering the era of precision black hole measurements."

A team of scientists led by Fabian used the Japanese X-ray observatory, Suzaka, to make a remarkable set of observations that clocked the speed of the black hole's spin rate, and evidence for a wall of X-ray light pulled back and flattened by gravity.

A second team, led by James Reeves of NASA's Goddard Space Flight Center in Greenbelt, Maryland, produced the first accurate measurement of the angle of a disk of material swirling around a black hole.

"The broad iron K line is our ticket to view matter and energy very close to a black hole," said Reeves. "Only by probing the extremes of gravity will we find flaws, if any, in Einstein's theories."

Fabian's team observed the galaxy called MCG-6-30-15, and found the black hole is spinning rapidly. They also saw evidence that X-rays emitted and trying to escape near its edge are bent back into the disk of material flowing inwards. Reeves' team observed a different galaxy, called MCG-5-23-16, and found the accretion disk is angled at 45 degrees with respect to our line of sight, the first time such detail has ever been determined.

http://www.nasa.gov/centers/goddard/news/topstory/2006/spinning_blackhole.html

http://www.jaxa.jp/projects/sat/astro_e2/index_e.html

http://arxiv.org/abs/astro-ph/0610434

http://arxiv.org/abs/astro-ph/0610436

■ *Below: 5 October 2006. An artist's impression of a black hole and its surrounding accretion disk. The Suzaku satellite is sensitive to key X-ray energies that enable scientists to discern properties of a black hole. Some X-rays are emitted at the edge of the black hole; others are reflected off of dust clouds several light-years away. Suzaku provides the most complete picture of this black hole activity, and when combined with optical and radio observations, even more can be learned. Image courtesy NASA/GSFC.*

26 October 2006

Latest Views of the V838 Monocerotis Light Echo from Hubble

The 2002 eruption of the star now known as V838 Monocerotis (V838 Mon), has been followed closely by the Hubble Space Telescope. Like a very bright flashbulb, the erupting star lit up previously dark clouds of dust, creating a spectacular "light echo". The illumination spreads at the speed of light, creating an ever-changing glimpse of the dusty environment surrounding V838 Mon. Each time Hubble takes an image of the region, it sees a new "thin section" of the surrounding dust veil.

The newest images reveal elegant swirls and eddies that would make a classical painter proud. The artwork etched on the sky is probably due to the interaction between the dust and magnetic fields that pervades the space between stars.

While the sheer beauty of the images is not diminished, images are taken for scientific purposes, and Hubble's Advanced Camera for Surveys used a polarizing filter to help astronomers use a technique for measuring the distance to V838 Mon. This placed a lower limit on the distance at 20,000 light-years. Further studies of the spectro-photometric parallax and other factors lead to a distance in excess of 30,000 light-years. This places the star in the outer part of the disk of our Galaxy, making the original outburst, which was visible in amateur telescopes on Earth, one of the brightest stars in the Milky Way.

Astronomers have yet to determine the exact cause for the eruption. U. Munari of the Padova Observatory, Italy, and colleagues suggest the eruption was caused by a thermonuclear eruption in the outer shell of a very massive O-type star.
http://www.spacetelescope.org/news/html/heic0617.html
http://arxiv.org/abs/astro-ph/0608222
http://arxiv.org/abs/astro-ph/0501604

26 October 2006

NASA's Spitzer Space Telescope Peels Back Layers of Supernova

Cassiopeia A is known as a fairly well-ordered supernova explosion. When the original massive star (15 to 20 times the mass of the Sun) erupted, the various layers inside the star were peeled back in the explosion, exposing heavier and heavier elements. Many of these elements have been observed in the Cassiopeia A supernova remnant, and now the Spitzer Space Telescope has filled in some vital gaps, confirming that the star erupted in a fairly uniform fashion.

"Now we can better reconstruct how the star exploded," said William Reach of NASA's Spitzer Science Center, Pasadena, California. "It seems that most of the star's original layers flew outward in successive order, but at different average speeds depending on where they started."

■ **Below**: 26 October 2006. These stunning images reveal changes in the light echo surrounding the star V838 Monocerotis between November 2005 and September 2006. They were taken by the Hubble Space Telescope's Advanced Camera for Surveys. The artist-like waves in the dust clouds lit up by the eruption on the star could possibly be produced by magnetic fields that fill interstellar space. Image courtesy NASA, ESA and H. Bond (STScI).

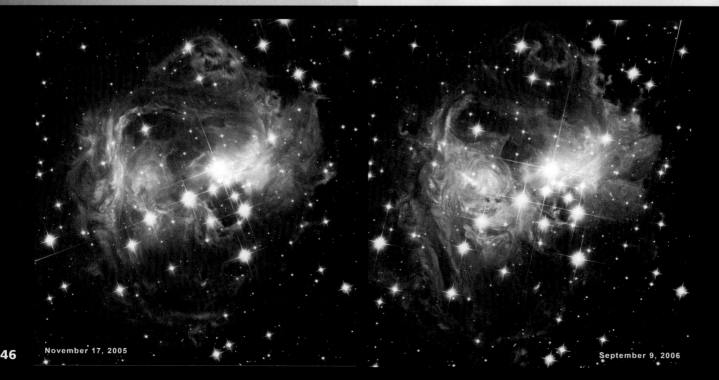

November 17, 2005

September 9, 2006

Each layer of the star expands outwards and one by one rams into the shockwave from the explosion. The fastest material that reached the shockwave first has had plenty of time to heat up, emitting X-ray and visible light. These components have been well observed before.

What Spitzer has found is infrared radiation from slower components that are just striking the shockwave and have not yet heated up significantly. These previously missing chunks are associated with a middle layer of elements from the original star such as neon, oxygen, and aluminum.

"Spitzer has essentially found key missing pieces of the Cassiopeia A puzzle," said Jessica Ennis of the University of Minnesota, Minneapolis. Leading the research, Lawrence Rudnick, also of the University of Minnesota, added, "We've found new bits of the 'onion' layers that had not been seen before. This tells us that the star's explosion was not chaotic enough to stir its remains into one big pile of mush."

http://www.spitzer.caltech.edu/Media/releases/ssc2006-19/release.shtml

http://arxiv.org/abs/astro-ph/0610838

NOVEMBER 2006

6 November 2006

Monster Stellar Flare Seen by NASA Scientists Dwarfs All Others

A massive flare erupted in December 2005 and was detected by the Swift gamma-ray observatory. Not only was the flare far larger than any seen on the Sun, but proves that Swift can do double duty, detecting stellar flares as well as gamma-ray bursts. The huge flare occurred on the main component of the close double star, II Pegasi, located 135 light-years from Earth, and weighs 0.8 times the mass of the Sun. Its companion is 0.4 solar masses.

"Swift was built to catch gamma-ray bursts, but we can use its speed to catch supernovae and now stellar flares," said Swift Project Scientist Neil Gehrels of NASA Goddard. "We can't predict when a flare will happen, but Swift can react quickly once it senses an event."

Soon after the flare generated a potential gamma-ray burst detection from its Burst Alert Telescope, scientists saw the X-ray Telescope on Swift become overwhelmed, and quickly realized this was a different kind of event. The flare released the energy equivalent of 50 million trillion atomic bombs and lasted for several hours. Flares on the Sun last for a few minutes. Had the Sun produced a flare like this, it would have overwhelmed Earth's atmosphere, with potentially catastrophic results.

"The flare was so powerful that, at first, we thought it was a star explosion," said Dr. Rachel Osten of University of Maryland. "We know much about solar flares on the Sun, but these are samples from just one star. This II Pegasi event was our first opportunity to study details of another star's flaring as if it were as close as our Sun," she added.

http://www.nasa.gov/mission_pages/swift/bursts/monster_flare.html

http://arxiv.org/abs/astro-ph/0609205

■ **Below:** *26 October 2006. The Spitzer Space Telescope shows the supernova remnant, Cassiopeia A. The representative color image shows the forward shock as a faint blue glow, while green, yellow and red reveal material ejected in the explosion and heated by a slower shock wave, called the reverse shock wave. The picture was taken by Spitzer's infrared array camera and is a composite of 3.6-micron light (blue); 4.5-micron light (green); and 8.0-micron light (red). Image courtesy NASA/JPL-Caltech/L. Rudnick (University of Minnesota).*

7 November 2006

VLT Shows Milky Way's Neighboring Galaxies Have Different History

The current cosmological model of the hierarchical formation of galaxies has been tested by observations from the Very Large Telescope (VLT), and failed, much to the surprise of astronomers.

If small galaxies formed first following the Big Bang and later merged to form larger galaxies, as current models predict, then astronomers expected dwarf galaxies orbiting our galaxy to contain stars with a lower abundance of heavier elements (referred to as metals). The reason is because dwarf galaxies, which formed in the early universe, largely contained hydrogen and helium and few metals. Nearly all other chemical elements have been synthesized in stars or in supernova explosions.

The VLT measured the iron concentration of over 2,000 stars in four nearby dwarf galaxies using FLAMES, the Fiber Large Array Multi-Element Spectrograph. FLAMES can simultaneously acquire 130 spectra of objects across a 25-arcminute field of view, and is ideally suited for this kind of survey.

The dwarf galaxies showed similar composition to stars in the galactic halo, but a key difference was the lack of metal-poor stars that are seen in the Milky Way. "The chemistry we see in the stars in these dwarf galaxies is just not consistent with current cosmological models," said Amina Helmi of the Kapteyn Astronomical Institute in Groningen, The Netherlands. "Our results rule out any merging of the nearby dwarf galaxies as a mechanism for building up the Galactic halo, even in the early history of the universe," she added.

■ *Right: 7 November 2006. Distribution of the iron content (on logarithmic scale) in four neighbouring dwarf galaxies of the Milky Way (Sculptor, Sextans, Fornax, and Carina), shown as relative fraction, as derived from the FLAMES/VLT observations. There is a great diversity from system to system, which reflects their widely different star formation and chemical enrichment histories. There is, however, a common denominator: a dearth of stars with very low amount of iron. Image courtesy European Southern Observatory.*

The observations are part of a large program called Dwarf galaxies Abundances and Radial-velocities Team (DART) involving scientists from nine countries.

http://www.eso.org/outreach/press-rel/pr-2006/pr-41-06.html

http://arxiv.org/abs/astro-ph/0611420

7 November 2006

Gravity Helps SDSS-II Reveal a Brilliant Jewel of the Early Universe

The Sloan Digital Sky Survey (SDSS-II) has discovered the brightest image of a galaxy ever recorded from the early universe. The galaxy has been distorted by an intervening cluster, producing a gravitationally lensed, arc-shaped, image of the more distant galaxy.

Sahar Allam of Fermilab discovered the galaxy during a long evening of searching for merging galaxies among 70,000 images from the survey. At 8 pm she found the arc, so dubbed it the "8 o'clock arc" The arc was actually four lensed images of the galaxy at a distance of 11.2 billion light-years, confirmed by a spectrum taken by the 3.5 meter telescope at Apache Point, New Mexico. Even the largest telescopes on Earth cannot study details in such distant galaxies because they are too faint. This discovery allows a unique view, thanks to the gravitational lens. They magnify distant galaxies, and this new galaxy is three times brighter than the previous record holder.

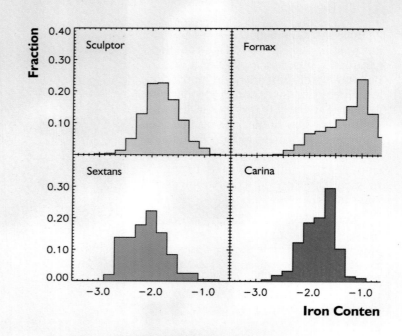

1 - A Year in News and Pictures

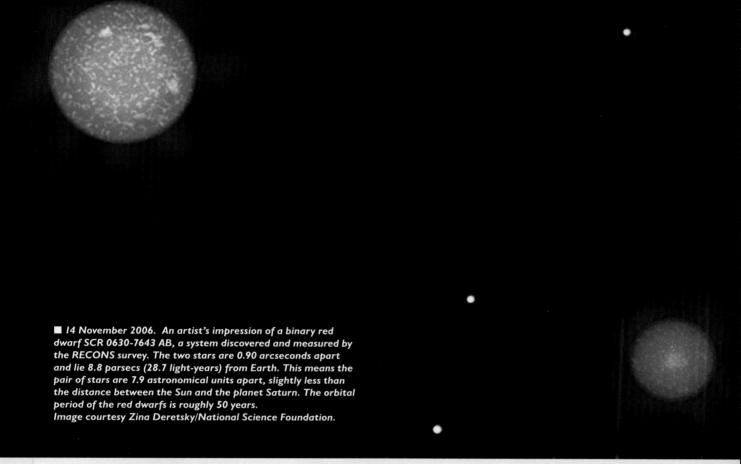

■ *14 November 2006. An artist's impression of a binary red dwarf SCR 0630-7643 AB, a system discovered and measured by the RECONS survey. The two stars are 0.90 arcseconds apart and lie 8.8 parsecs (28.7 light-years) from Earth. This means the pair of stars are 7.9 astronomical units apart, slightly less than the distance between the Sun and the planet Saturn. The orbital period of the red dwarfs is roughly 50 years.*
Image courtesy Zina Deretsky/National Science Foundation.

"Lots of SDSS science emerges from the overwhelming statistics of enormous samples," said Allam. "But this is a completely different way of mining the SDSS for scientific discoveries. If you're willing to look through tens of thousands of rocks, every so often you find a jewel."

http://home.fnal.gov/~sallam/8OClockArc/
http://arxiv.org/abs/astro-ph/0611138

14 November 2006

Twenty New Stars in the Neighborhood

The Research Consortium on Nearby Stars (RECONS) announced the discovery of twenty new stars within 33 light-years of Earth. All the new stars are red dwarfs, and give a total population of objects within 10 parsecs (32.6 light-years) at 348. Sixty-nine percent of these are red dwarfs, and may reflect the stellar population across the Milky Way if our region of the galaxy is typical.

"Our goal is to help complete the census of our local neighborhood and provide some statistical insights about the demographics of stars in our galaxy – their masses, their evolutionary states, and the frequency of multiple star systems," says RECONS Project Director Todd Henry of Georgia State University in Atlanta. "Due to their proximity, these systems are also excellent targets for exoplanet searches, and ultimately, for astrobiological studies of whether any planets that are found could support life."

The distances were measured by trigonometric parallax using the 0.9-meter telescope at the Cerro Tololo InterAmerican Observatory in Chile. Trigonometric parallax uses the orbital motion of the Earth and its effect on the position of nearby stars relative to background stars to measure distance. Two opposing sides of Earth's orbit provide a large baseline from which the angular shift of a star can be measured.

http://www.noao.edu/outreach/press/pr06/pr0614.html
http://www.chara.gsu.edu/RECONS/
http://arxiv.org/abs/astro-ph/0608230

53" x 53" IRCS 52mas camera

20 November 2006

Subaru Telescope Improves its Eyesight by a Factor of Ten

The Japanese Subaru telescope, located on Mauna Kea, Hawaii, has upgraded to a new adaptive optics system, following in the newly worn footsteps of the groundbreaking work at the Lick, Keck, and Gemini observatories. The system allows a factor ten improvement in resolution and greater contrast. The laser guide star system, combined with the 188-element adaptive optics device, shines a laser star into a layer of atmosphere, and its disturbances in the image caused by turbulence is counterbalanced by very fast physical manipulations of the shape of the 130-mm diameter adaptive mirror.

The system has been demonstrated by comparing an image of the Trapezium region of the Orion Nebula obtained at first light in 1999 with a new image taken with the adaptive optics system. Subaru achieved typical resolution of 0.6 arcseconds before the upgrade, and now achieves 0.06 arcseconds. While some of the instrumentation has been developed elsewhere, Japan produced its own of solid-state laser and optical fiber technology, representing original contributions to the rapidly developing field.

With the record breaking distant galaxies Subaru has been detecting, the addition of the adaptive optics system will sustain the Subaru telescope at the forefront of astronomical research for many years to come.

http://www.subarutelescope.org/Pressrelease/2006/11/20/index.html

http://subarutelescope.org/Introduction/instrument/AO.html

■ **Above:** *20 November 2006. A comparison of the "before" and "after" image of the Trapezium region in the Orion Nebula obtained with and without adaptive optics. The Subaru Telescope acquired these images using a new adaptive optics system that gives it a ten-fold increase in resolution. The image on the right, obtained in 1999, has a resolution of 0.6 arcseconds. The image on the left has a resolution of 0.06 arcseconds and was obtained in October 2006 using adaptive optics. Image copyright © Subaru Telescope, NAOJ. All rights reserved.*

27 November 2006

Integral Catches a New Erupting Black Hole

Integral, the European Space Agency's gamma-ray observatory, detected an unusual gamma-ray burst on 17 September 2006. It was the first detection of a four week intensive study of the central region of the Milky Way. The burst, instead of fading rapidly, continued to rise for a few days. XMM Newton, Swift, and the Chandra X-ray telescopes contributed observations following a worldwide alert sent out from the Integral mission control.

"It was only after a week that we could see the shape of the light curve and realized what a rare event we had observed," said Roland Walter, an astronomer at the Integral Science Data Centre (ISDC), Versoix, Switzerland.

The light curve suggests astronomers have detected a star-black hole binary, and the burst came from the collapse of the accretion disk as it became unstable and fell into the black hole. The predicted rate of such events is once every few years, so this detection was a unique chance to observe the details of what happened. The reasons for accretion disk collapse are not well understood.

http://www.esa.int/esaSC/SEM8SDANMUE_index_0.html

http://isdc.unige.ch/Science/news/061123/

DECEMBER 2006

6 December 2006

Do Galaxies Follow Darwinian Evolution? VLT Survey Provides New Insight into Formation of Galaxies

The Very Large Telescope (VLT) has spent some of its valuable observing time over the past three years studying over 6,500 galaxies at varying distances from Earth, all the way out to 9 billion light-years. Looking at the huge atlas of galaxies generated by the VLT, astronomers have come to some interesting conclusions relating to a vibrant topic in current research – the evolution of galaxies.

■ 27 November 2006. An artist's impression of an X-ray nova is shown to illustrate an event detected by the Integral gamma-ray observatory. The rare event suggests an accretion disk around a neutron star or black hole collapsed, emitting a burst of gamma-rays. In this illustration, the compact object on the right is either a neutron star or a black hole. Material from a companion star spirals onto the accretion disk. The gas swirls in a disk around the compact object at very high velocity (close to the speed of light) and emits X-rays. Image courtesy ESA.

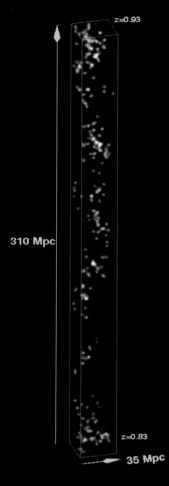

z=0.93

310 Mpc

z=0.83

35 Mpc

"Our results indicate that environment is a key player in galaxy evolution, but there's no simple answer to the 'nature versus nurture' problem in galaxy evolution," said Olivier Le Fevre from the Laboratoire d'Astrophysique de Marseille, France.

The three-dimensional atlas of the universe, spanning 9 billion light-years, enabled astronomers to study the color, luminosity, and density of galaxy cluster in which each galaxy resides, and to come to their conclusions. The overall color of the galaxy indicates the level of star formation, with redder galaxies having less star formation than bluer ones.

The redder galaxies tend to be part of denser galaxy clusters. This suggests that members of galaxy clusters have their ability to form stars stifled more quickly compared to more isolated galaxies. The discovery of this so-called "color-density" relationship implies a galaxy's primordial formation is not the only factor affecting future evolution, but environment is also a strong factor.

The results come from the VIMOS (Visible Multi-Object Spectrograph) Deep Survey performed with one of the European Southern Observatory's four 8.2-meter unit telescopes on Cerro Paranal, Chile. VIMOS can acquire spectra of 1,000 objects on a single field of view simultaneously. The Deep Survey continues to scan sixteen square degrees of sky and is producing a goldmine of data.

http://www.eso.org/outreach/press-rel/pr-2006/pr-45-06.html

http://arxiv.org/abs/astro-ph/0603202

http://arxiv.org/abs/astro-ph/0602329

http://arxiv.org/abs/astro-ph/0701273

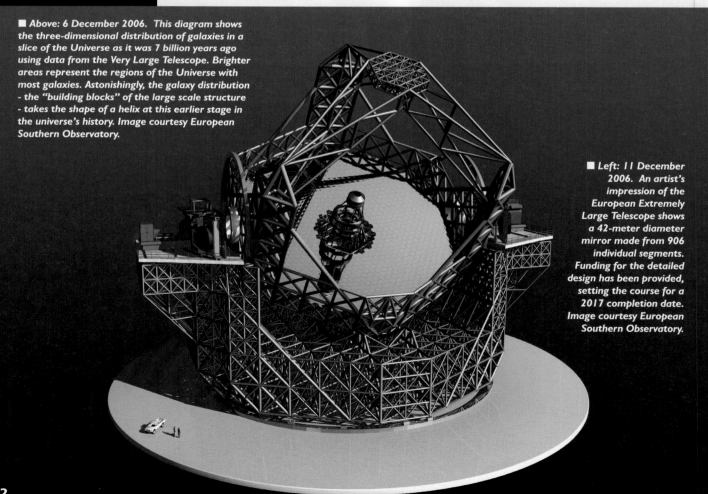

■ Above: 6 December 2006. This diagram shows the three-dimensional distribution of galaxies in a slice of the Universe as it was 7 billion years ago using data from the Very Large Telescope. Brighter areas represent the regions of the Universe with most galaxies. Astonishingly, the galaxy distribution - the "building blocks" of the large scale structure - takes the shape of a helix at this earlier stage in the universe's history. Image courtesy European Southern Observatory.

■ Left: 11 December 2006. An artist's impression of the European Extremely Large Telescope shows a 42-meter diameter mirror made from 906 individual segments. Funding for the detailed design has been provided, setting the course for a 2017 completion date. Image courtesy European Southern Observatory.

■ 11 December 2006. The star cluster Pismis 24 lies in the core of the large emission nebula NGC 6357 that extends one degree on the sky. It lies 8,000 light-years away in the direction of the Scorpius constellation. The cluster contains three of the top 25 most massive stars known in our galaxy. Pismis 24-1, once thought to be a single star, has been shown to contain three stars, each at least 50 solar masses. The intense ultraviolet radiation from these massive hot stars heats the gas surrounding the cluster and creates a bubble in NGC 6357. Image courtesy NASA, ESA and Jesús Maíz Apellániz (Instituto de Astrofisica de Andalucía, Spain), with thanks to Davide De Martin (ESA/Hubble).

11 December, 2006

Heavyweight Stars Light Up Nebula NGC 6357

In the star cluster Pismis 24, a star previously thought to be one of the most massive stars known has turned out to be three stars. The original suspicion was that the star, Pismis 24-1, weighed between 200 and 300 times the mass of the Sun, well above the theoretical limit for single stars. It lies 8,000 light-years away in the direction of the constellation of Scorpius. The Hubble Space Telescope was able to resolve the star into a pair, a binary star. Ground-based telescopes further detected a third star orbiting the primary using spectroscopic observations. The main pair is estimated to be 100 solar masses, among some of the heaviest known stars. This discovery does resolve the problem of a more massive star, because the theoretical upper limit for a single star is about 150 solar masses.

The cluster of stars, associated with a spectacular emission nebula, also contains a third massive star, Pismis 24-17. Its mass is estimated at 100 times the mass of the Sun. Massive stars have short lifetimes of just a few million years and finding three such massive stars in one cluster is rare. The observations were led by Jesús Maíz Apellániz of the Astrophysics Institite at Andalucía, Spain.

Observations of the cluster made by Junfeng Wang of Pennsylvania State University, and colleagues, using the Chandra X-ray Observatory estimate that the Pismis 24 cluster contains about 10,000 stars, making it richer than the Orion Nebula cluster. The astronomers also detected an X-ray emitting infrared source which they suggest could be a heavily-obscured massive Young Stellar Object (YSO) or alternatively, a Wolf-Rayet star. Further observations of this rich region of recent star formation will yield some more answers.

http://arxiv.org/abs/astro-ph/0609304
http://www.spacetelescope.org/news/html/heic0619.html

■ Opposite page: 15 December 2006. Three artist's impressions showing how two massive stars form together within a larger gas cloud. Observations of the multiple-star system L1551 IRS5 by the Very Large Array (VLA) radio telescope support the idea that such massive stars are formed together in the same location. Top: This picture shows a large disk-like rotating cloud of gas and dust. Middle: Here, two smaller disks of gas and dust fragment from the large disk and begin to condense into protostars. Each protostar sports its own disk, aligned with the main disk, and shoots "jets" of material outward from the poles. Bottom: In this view, a third smaller disk and protostar have joined the system, either through the same fragmentation process or by being captured gravitationally by the larger protostars. Images courtesy Bill Saxton, NRAO/AUI/NSF.

11 December 2006

ESO Council Gives Green Light to Detailed Study of the European Extremely Large Telescope

Detailed studies for the construction of the Extremely Large Telescope, an optical and infrared telescope with an aperture of 42-meters, have been given the go-head by the Council of the European Southern Observatory. The design study will result in the possibility of construction beginning in 2009 at the earliest.

"At the end of the three year Final Design Study, we will know exactly how everything is going to be built, including a detailed costing," said Catherine Cesarsky, Director General of ESO. "We then hope to start construction and have it ready by 2017, when we can install instruments and use it!"

A unique five-mirror design includes a huge 42-meter aperture, achieved using 906 hexagonal elements, each about 1.45-meters across. A 6-meter secondary mirror will feed starlight to a 4.2-meter tertiary mirror, and then into an adaptive optics system to remove the blurring effects of Earth's atmosphere. This is performed by a deformable 2.5-meter mirror and is manipulated by 5,000 actuators. A fifth 2.7-meter mirror makes final image corrections.

The current name is the European Extremely Large Telescope (E-ELT) and it will be a hundred times more sensitive than any existing ground-based telescope.

http://www.eso.org/projects/e-elt/

15 December 2006

How Do Multiple-Star Systems Form? VLA Study Reveals "Smoking Gun"

The proto-stellar system called L1551 IRS5, located about 450 light-years away in the direction of Taurus, has revealed new details about massive star formation to observers using the Very Large Array radio telescope in Socorro, New Mexico. Earlier observations using half of the VLA's 27 antennas revealed a pair of stars surrounded by a cloud of dust and gas. The new observations, using the full resolution of the entire array plus one segment of the Very Long Baseline Array, provide strong evidence that two massive stars have formed out of the same cloud.

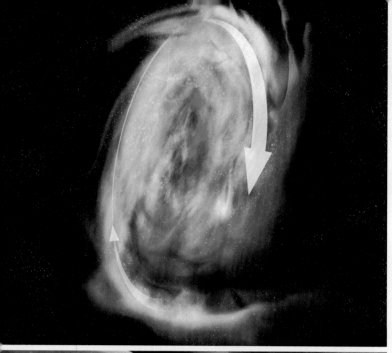

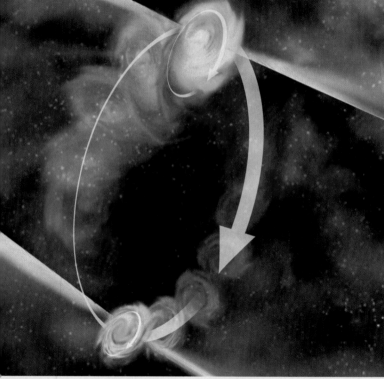

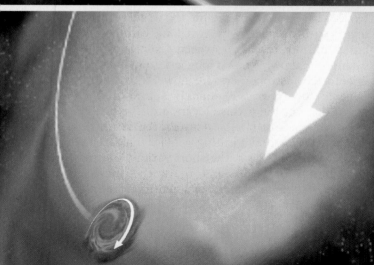

Competing theoretical models of how multiple star systems form suggest they are either from a larger parent disk and fragment in-situ, or alternatively, they form individually and later are captured.

"We have a very firm indication that two of these protostars and their dust disks formed from the same larger disk-like cloud, then broke out from it in a fragmentation process. That strongly supports one theoretical model for how multiple-star systems are formed," said Jeremy Lin of the Institute of Astronomy & Astrophysics, Academia Sinica, Taiwan.

The new observations show that the disks surrounding both protostars are aligned with each other, and also aligned with a larger disk of material in which the new stars are embedded. Lim, along with colleague, Shigehisa Takakuwa, discovered a third proto-star. However, its disk is not aligned with the other pair. This additional object suggests there may be more than one way to form multiple star systems, since the third object could have been formed elsewhere and captured later.

Previous observations by the Hubble Space Telescope (HST), the SubMillimeter Array (SMA), and XMM Newton have targeted this heavily obscured system. XMM Newton discovered the X-ray emission from a jet emanating from the proto-stellar system and striking the circumstellar gas. Future telescopes such as the ALMA array and the Square Kilometer Array will allow astronomers to see even more detail inside this and other multiple star forming regions.

http://www.nrao.edu/pr/2006/multidisk/
http://arxiv.org/abs/astro-ph/0110112
(XMM Newton)
http://arxiv.org/abs/astro-ph/0212074
http://arxiv.org/abs/astro-ph/0404346 (SMA)
http://arxiv.org/abs/astro-ph/0503469 (HST)

18 December 2006

Spitzer Space Telescope Picks Up Glow of Universe's First Objects

The very first objects that formed in the universe following the Big Bang have their light shifted well into the infrared region of the electromagnetic spectrum. New observations from the Spitzer Space Telescope suggest it may have detected the first tantalizing glimpse of light from these long sought after primordial objects.

"We are pushing our telescopes to the limit and are tantalizingly close to getting a clear picture

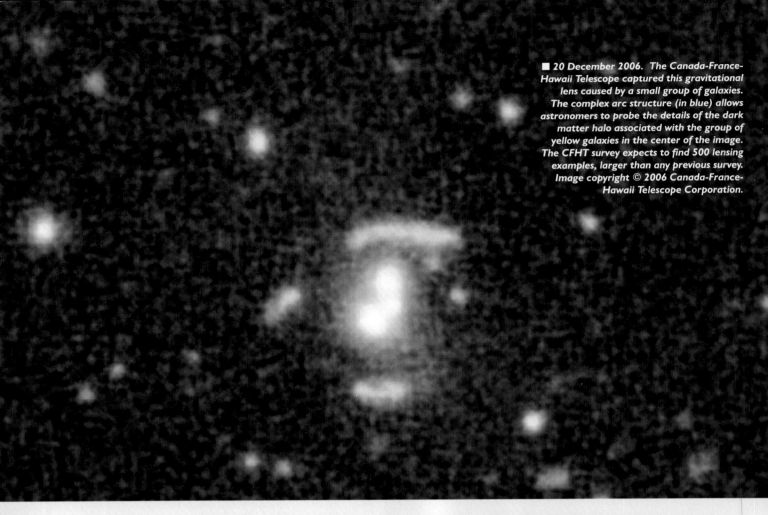

of the very first collections of objects," said Dr. Alexander Kashlinsky of NASA's Goddard Space Flight Center (GSFC) in Greenbelt, Maryland. The observations required hundreds of hours of exposure time and cover five regions of the sky.

Kashlinksy and colleagues studied the cosmic infrared background with Spitzer by removing all the foreground stars and galaxies. The remaining light in the image is diffuse, and is thought to reveal clustering of some objects. Light may come from activity around some of the universe's first black holes, or be the result of starlight from some of the earliest and most massive first generation of stars.

"There's ongoing debate about what the first objects were and how galaxies formed," said Dr. Harvey Moseley of GSFC.

The first stars ended the Dark Ages, the era after the universe became transparent to radiation following the Big Bang and before the first stars lit up. The original emission of light was in the ultraviolet and optical, and due to the expansion of the universe during the intervening 13 billion years or so, its wavelength has been stretched into the infrared. Future large infrared telescopes, such as SOFIA, and the James Webb Space Telescope, will be targeting this primordial glow to answer some of the basic questions we have about the origin of the universe.

http://www.spitzer.caltech.edu/Media/releases/ssc2006-22/release.shtml

http://arxiv.org/abs/astro-ph/0612445

http://arxiv.org/abs/astro-ph/0612447

20 December 2006

Two Cosmic Bursts Upset Tidy Association Between Long Gamma-ray Bursts And Supernovas

Dozens of astronomers around the world teamed together to discover why two long gamma-ray bursts did not reveal an expected supernova in each host galaxy. When a long gamma-ray burst is detected, astronomers are now quite sure a Type Ic supernova will quickly appear in the host galaxy at the precise location of the burst. The long class of bursts signifies the death of a massive star and forming a black hole. The second type of GRB is the short variety, lasting less than 2 seconds, and mark the merger between two neutron stars, or a neutron star and black hole.

In May and June 2006, two gamma-ray bursts occurred that are challenging this simple picture. GRB 060505 and GRB 060614 were detected by the orbiting Swift observatory and produced bursts of 4 seconds and 102 seconds respectively. In the days that followed, the Danish 1.54-meter telescope on La Silla, Chile, the 10-meter Keck, and the 8-meter Gemini telescopes, as well as the Hubble Space Telescope, each observed the host galaxies, and no supernova was detected.

The long burst from GRB 060614 occurred in a galaxy 1.6 billion light-years away, and is referred to as a hybrid burst - a long burst without a supernova. It appeared to resemble a short burst, yet there are no ideas of how the progenitors of a short burst could produce a sustained long burst.

"We have lots of data on this event, have dedicated lots of observation time, and we just can't figure out what exploded," said Neil Gehrels of NASA's Goddard Space Flight Center in Greenbelt, Maryland and head of the Swift mission. "This is brand new territory; we have no theories to guide us," said Gehrels, commenting about the 102-second long burst.

"This burst was close enough to detect a supernova if it existed," said Avishay Gal-Yam of Caltech, "but even Hubble didn't see anything."

Astronomers are suspecting that a new type of black hole formation was observed, one in which all the material is swallowed by the black hole following a massive supernova explosion. Astronomers at Penn State University have shown the long burst appears to be a scaled-up version of a short burst, implying a link to black hole formation.

Joshua Bloom, of the University of California at Berkeley, who last year discovered evidence that linked short-lived gamma-ray bursts to the merger of neutron stars and black holes old said, "Instead of simplicity and clarity, we're seeing a rich diversity emerge - there are more ways than we thought for producing flashes of gamma-rays."

http://hubblesite.org/news/2006/55

http:// www.eso.org/outreach/press-rel/ pr-2006/pr-49-06.html

http://www.science.psu.edu/alert/ Fox12-2006.htm

http://arxiv.org/abs/astro-ph/0608313

http://arxiv.org/abs/astro-ph/0612238

(Note: Neil Gehrels and his colleague Peter Leonard contributed a review chapter entitled 'The Most Powerful Explosions in the Universe… Gamma-Ray Burst Discoveries with the Swift Mission' in last year's volume, State of the Universe 2007.)

20 December 2006

New Gravitational Lensing by Groups of Galaxies

Gravitational lensing of distant galaxies by massive foreground galaxy clusters has been studied closely, especially by the Hubble Space Telescope. However, a new ground-based survey is uncovering many lensed systems around smaller groups of galaxies.

The new discoveries come from the Canada-France-Hawaii Telescope (CFHT) on Mauna Kea, the same telescope that obtained the first CCD image of a gravitational lens in January 1987. The CFHT Legacy survey has dedicated 500 nights over five years to a large scale survey that will eventually cover one percent of the sky. It will be completed by 2008.

The survey is being conducted using the 340 megapixel camera, MegaCam, which can cover one square degree in a single prime focus image on CFHT. One major step forward is the automated search for lensed structures. Remi Cabanac of the CFHT, and colleagues, expect to find over 500 lensed structures in the Legacy survey, outnumbering all existing galaxy lens samples currently known. The best examples will be followed up by Hubble Space Telescope observations.

The lensed galaxies provide valuable data on the dark matter distribution in the intervening groups of galaxies, critical in the quest to understand large scale structures in the universe. Lensing caused by these intermediate scale groups of galaxies can provide insight to the evolution of structures over time, since there is a larger statistical sample of lensed galaxies over a wide range of redshifts than any previously available.

http://www.cfht.hawaii.edu/News/ StrongLensing/

http://arxiv.org/abs/astro-ph/0610362

JANUARY 2007

4 January 2007

X-Ray Evidence Supports Possible New Class of Supernova

The two largest orbiting X-ray observatories, NASA's Chandra and ESA's XMM Newton, have uncovered evidence for a new class of supernova explosion. The evidence comes from data obtained from two supernova remnants, DEM L238 and DEM L249, until now thought to be the product of normal Type Ia supernovas, caused by the explosion of white dwarf stars. While the tell-tale large signal of iron is present in both remnants, the hot gas is brighter and denser than expected for a Type Ia event. The suggestion is that the progenitors come from a population of fast evolving stars.

The observed X-ray data has been successfully matched to computer simulations of a white dwarf exploding in a dense environment. Quickly evolving massive binary stars produce huge stellar winds and high mass loss rates, producing the dense environment around a star. Once one star reaches the white dwarf stage, fast mass loss from its companion can dump material onto the white dwarf producing a supernova.

"We know that the more massive a star is, the shorter its lifetime," said Kazimierz Borkowski of North Carolina State University, Raleigh. "If such a star could also begin to pull matter from its companion at an early stage, then this star would have a much shorter fuse and explode in only about 100 million years – much less than other Type Ia supernovas."

These "prompt" Type Ia events may be visible at greater distances than previously supposed. Type Ia supernovas are used as standard candles, because their explosions are consistent in their light output. If the first stars exploded more quickly, as these new observations suggest, it would allow astronomers to probe earlier stages of the universe's history. Now astronomers are searching for other examples, and one prime candidate is Kepler's supernova of 1604. (See also 9 January story.)

http://chandra.harvard.edu/photo/2007/deml238/

http://arxiv.org/abs/astro-ph/0608297

5 January 2007

Google Joins Large Synoptic Survey Telescope Project

When it comes to organizing massive quantities of data and allowing easy access, Google has it covered. So it comes with little surprise that Google is teaming up with the group of nineteen universities on the Large Synoptic Survey Telescope (LSST) project. Everything about the LSST is huge, especially the data handling aspect.

Aside from the 8.4-meter mirror, LSST will survey the entire sky every three nights,

■ *Right: 4 January 2007. Two supernova remnants observed by the Chandra X-ray Observatory, called DEM L238 and DEM L249, are strong candidates for a new type of supernova explosion. Shown here as a combination of X-ray (blue) and optical (white) images, the remnants are similar to Type Ia supernova remnants. The green color in the center of DEM L238 indicates an overabundant iron-rich central region. This suggests the progenitor stars were more massive and were very young when they underwent a thermonuclear explosion compared with typical Type Ia explosions. These "prompt" Type Ia events are predicted to be visible at far greater distances than previously supposed. Images courtesy NASA/ CXC/NCSU/K.Borkowski – X-ray; NOAO/CTIO/ MCELS – Optical.*

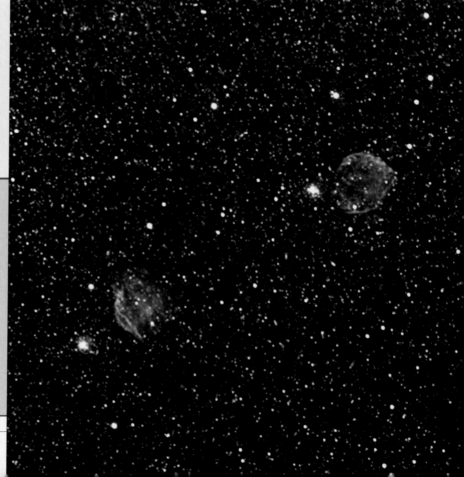

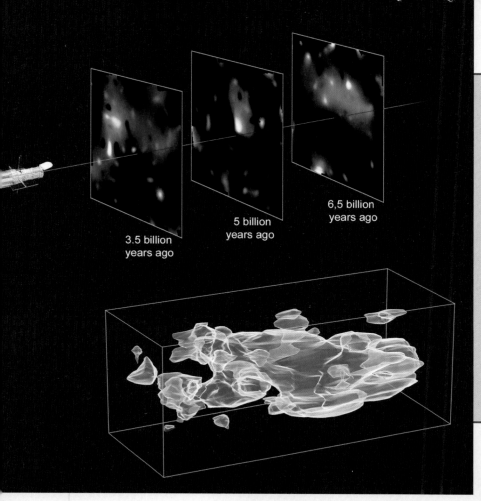

3.5 billion
years ago

5 billion
years ago

6,5 billion
years ago

producing thirty terabytes (30,000 Gb) of data. Managing this continuously changing data and providing real-time access has some parallels to managing access to the ever-changing internet. The goal is to allow amateur and professional astronomers alike to browse this new versatile view of the night sky like never before.

LSST Director, J. Anthony Tyson, a professor at the University of California at Davis, notes, "Partnering with Google will significantly enhance our ability to convert LSST data to knowledge. LSST will change the way we observe the universe by mapping the visible sky deeply, rapidly, and continuously."

http://www.lsst.org/News/google.shtml

7 January 2007

Lessons from a Star Disk

Astronomers from the University of California at Berkeley have discovered some remarkably porous dust around the nearby star, AU Microscopii (AU Mic), filling in a missing link between the presence of dusty disks around stars and the formation of planet-sized objects.

The porosity of the dust was measured using a polarizing filter on the Advanced Camera for Surveys aboard the Hubble Space Telescope. AU Mic, which lies 32 light-years from Earth, is close enough for detailed observations of its disk. The original dusty disk was discovered by one of the researchers in this study, Paul Kalas, in 2004, and it's the nearest known system to Earth. The brightness of scattered light at different polarizations allows the porosity to be calculated. The dust is found to be 97 percent porous, similar to champagne powder snow on Earth. Only 3 percent is ice.

Such fine dust is easily blown away from the star by its stellar wind, so the dust must be constantly replenished by collisions of larger bodies in the inner disk. These larger bodies are thought to be 10-cm to 20-cm sized snowballs, and glancing collisions produce puffs of ice particles about as fine as smoke particles.

"We have seen many seeds of planets and we have seen many planets, but how they go from one to the other is a mystery," said James Graham of the University of California at Berkeley. "These observations begin to help us fill in that gap."

The disk extends from about 40 astronomical units to several hundred, equivalent to the Kuiper Belt in our own solar system. Paul Kalas estimates the mass of the disk at about equivalent to one Moon mass. "If we ran our solar system clock back nearly 4.5 billion years, then the infant Kuiper Belt would probably look like AU Microscopii's birth ring," said Kalas.

The observations fit with a general theory of planet formation in which dust and gas coalesce to form larger bodies in the first 10 million years or so. While larger bodies form planet-sized objects, finer particles get blown out of the inner region, leaving a hole dominated by the larger objects, such as the one found in the AU Mic disk.

http://hubblesite.org/newscenter/archive/releases/2007/02/

http://www.berkeley.edu/news/media/releases/2007/01/08_dust.shtml

http://www.ifa.hawaii.edu/info/press-releases/AU_Mic2-24-04.html

http://arxiv.org/abs/astro-ph/0609332

(Note: See also the review chapter entitled 'Chips off the Planetary Block …Building Planetary Debris Disks' by the researchers quoted in this story, James Graham and Paul Kalas, in Chapter 6 of this book.)

■ *Below: 7 January 2007. This annotated image from the Hubble Space Telescope's Advanced Camera for Surveys reveals the almost edge-on dusty disk of AU Microscopii, a star located 32 light-years from Earth. The coronograph mode was used to block out light from the star, enabling the much fainter disk to be viewed. Observations of the polarization of light in the disk reveal the presence of fluffy dust particles. In order for these to be present, the porous dust particles must be replenished by repeated collisions of larger particles in the inner disk. Image courtesy NASA, ESA, J.R. Graham and P. Kalas (University of California, Berkeley), and B. Matthews (Hertzberg Institute of Astrophysics).*

7 January 2007

The First Direct Maps of the Three-Dimensional, Large-Scale Distribution of Mass in the Universe

A new result from the COSMOS survey (reviewed by Anton Koekemoer in last year's volume, State of the Universe 2007) has revealed the three dimensional distribution of dark matter for the first time. The breakthrough technique, developed by astronomers from the California Institute of Technology (CalTech), measured the shapes of half a million galaxies at varying distances. The subtle distortion in their shapes caused by weak gravitational lensing along the line of site allowed a reconstruction of the intervening dark matter spanning billions of light-years, using image slice techniques similar to tomography.

Astronomers are trying to "understand the background wallpaper of the galactic distribution to unravel the Dark Matter gravitational lensing effect," explained Richard Massey of CalTech.

The results were made possible by the large COSMOS survey. The Hubble Space Telescope made 575 images over a 2° field of view, a project

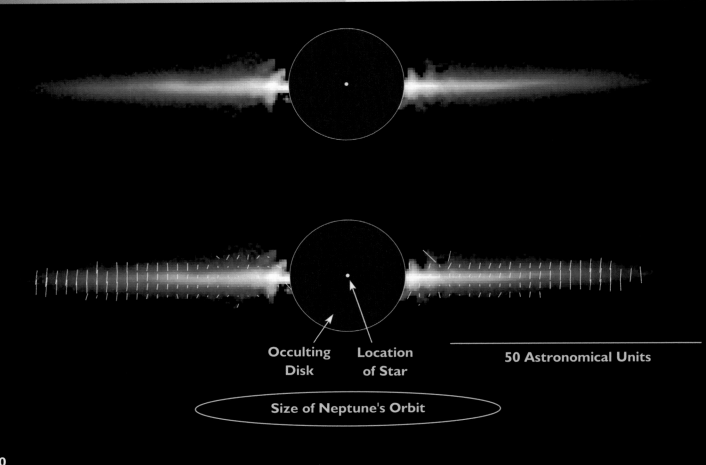

Occulting Disk

Location of Star

50 Astronomical Units

Size of Neptune's Orbit

that absorbed ten percent of Hubble's observing time over a two-year period.

The map stretches back to half the age of the universe, and reveals how the distribution of dark matter has become clumpier as gravity has done its work over time. It's a study of "paleocosmology, peeling back the layers of time," said Nick Scoville of CalTech.

"It's reassuring how well our map confirms the standard theories for structure formation," said Massey. "All visible matter lies within the scaffolding of the Dark Matter structure," he added.

The growth of the large scale structure has been predicted by numerical simulations for many years, and this first clear result is a major step forward for astronomers. The distortion amounts to only 1-2-percent of a galaxy's shape, and requires many galaxies to detect the effect. The weak lensing technique was developed in 2000, and the first results came out in 2003. These new results give time slices, and the method has become a fundamental tool for the cosmologist.

http://hubblesite.org/news/2007/01

http://arxiv.org/abs/astro-ph/0701594

7 January 2007

Astronomers Discover Red Giant Halo Orbiting Andromeda Galaxy

The Andromeda Galaxy is five times larger than previously thought, astounding astronomers who made the discovery of a halo of red giant stars orbiting 500,000 light-years from the center of the galaxy. Andromeda lies only 2.5 million light-years from Earth. The newly discovered stars appear to be the elusive halo sought by astronomers for a couple of decades.

"I am absolutely astounded by how big this halo is. As we looked farther and farther out, we kept finding stars that look like halo stars," said Puragra (Raja) Guhathakurta, professor of astronomy and astrophysics at the University of California at Santa Cruz (UCSC) and leader of the team making the discovery.

The new stars fit models of galaxy formation by revealing a lower percentage of heavy elements compared with other stars. These "metal-poor" stars (in astronomy, metals refer to any element heavier than hydrogen or helium) are probably the oldest stars in the Andromeda Galaxy. Galaxy formation theories based on our own Milky Way suggest these halo stars form first

and are pristine. The observations of Andromeda provide an excellent test of these theories.

"The physical size of this galaxy is really striking," said R. Michael Rich of UCLA. "The suburbs of M31 and the Milky Way are so extended that they nearly overlap in space, despite the great distance between these two galaxies. If the whole of M31 were bright enough to be visible to the naked eye, it would appear to be huge, larger in apparent size than the Big Dipper."

Red giants in M31 are hard to pick out from faint, nearby, foreground stars that are members of our own galaxy. Karoline Gilbert, a UCSC graduate student, developed the technique for separating these two distinct populations of stars using nearly half a dozen criteria, including photometry, radial velocity measurements, and spectroscopy.

"We focused on detecting red giant stars in the halo because they are bright enough for us to obtain spectra," Gilbert said. "There are assuredly other kinds of stars in Andromeda's halo, but they are just too faint for us to get spectra of them." The data were obtained using the 4-meter Mayall Telescope on Kitt Peak and the 10-meter Keck telescope in Hawaii.

http://www.keckobservatory.org/article.php?id=102

http://arxiv.org/abs/astro-ph/0605171

http://arxiv.org/abs/astro-ph/0502366

8 January 2007

The First Triple Quasar

At first sight, the appearance of three quasars in very close proximity suggests they may be the result of a gravitational lens. George Djorgovski of the California Institute of Technology thought so, until extensive searching for an intervening lensing galaxy came up blank, proving the trio must be the first triple quasar ever discovered.

The first and brightest component was discovered in 1989 by British astronomers, and the second quasar was assumed to be a gravitational lens. Observations from the European Southern Observatory's 8.2-meter VLT and one of the twin 10-meter Keck telescopes spotted the third quasar.

The research team, led by Djorgvski, performed extensive computer modeling, effectively ruling out the chance astronomers were seeing a gravitational lens effect. Astronomers had their first triple quasar. Differences in each quasar spectrum bolster the claim. The trio lies 10.5 billion light-years away, and occupies a space and time when galaxy interactions were frequent. Quasars are thought to be powered by material falling in to supermassive black holes.

"As more binary and triple quasar systems are discovered, science will have a new tool with which to understand how galaxies and supermassive black holes in the distant universe may have developed and changed over time," said Dr. Taft Armandroff, director of the W.M. Keck Observatory.

http://www.keckobservatory.org/article.php?id=98

http://www.eso.org/outreach/press-rel/pr-2007/pr-02-07.html

http://arxiv.org/abs/astro-ph/0701155

8 January 2007
Supergiant Goes to Extremes

Frequent outbursts of a hypergiant star in the constellation of Canis Major have come under the scrutiny of the W.M. Keck Telescope and the Hubble Space Telescope. The red supergiant star, VY Canis Majoris, is located 5,000 light-years away, and is one of the largest and most luminous evolved stars known. It is 500,000 times brighter,

and about 30 to 40 times more massive than the Sun. Its bloated surface would reach the orbit of Saturn if it were placed in our solar system. Outbursts are a regular occurrence, and it's been under watch for more than a century.

During an outburst, the star ejects material into space, and most models have assumed spherical symmetry, but the new observations show that it is "very complex." Images reveal a wide range of features, showing that eruptions shoot knots of material in different directions and at different speeds. Arcs of material reveal a history of repeated eruptions from different active regions on the star. Spectroscopy with both telescopes has unraveled the complex motions. The outermost material appears to have been ejected 1,000 years ago, while the most recent could have been thrown off as little as 50 years ago.

"With these observations, we have a complete picture of the motions and directions of the outflows, and their spatial distribution, which confirms their origin from eruptions at different times from separate regions on the star," said Roberta Humphreys of the University of Minnesota.

VY Canis Majoris is nearing the end of its life. Prior to exploding as a supernova, red supergiant stars shed mass, and this process is well underway now, with nearly half of the star's mass already ejected.

http://hubblesite.org/newscenter/archive/releases/2007/03/

http://arxiv.org/abs/astro-ph/0702717
http://arxiv.org/abs/astro-ph/0702718
http://arxiv.org/abs/astro-ph/0610433

■ *Below: 8 January 2007. A pair of images of the hypergiant star VY Canis Majoris from the Hubble Space Telescope. The left image is combined from two photographs taken by the Wide Field and Planetary Camera 2 in 1999 and 2005. The second image (right) was taken in 2003 by the Advanced Camera for Surveys through polarizing filters. VY Canis Majoris is located 5,000 light-years away, and is a huge star, one of the largest known. It shines with 500,000 times the brilliance of the Sun. The material around the star shows how chaotic various outbursts by the star have been in the past. The random arcs probably came from localized eruptions in active regions on the star's surface. Image courtesy NASA, ESA, and R. Humphreys (University of Minnesota).*

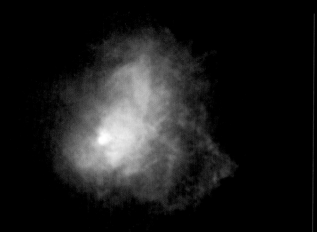

Visible Light
WFPC2

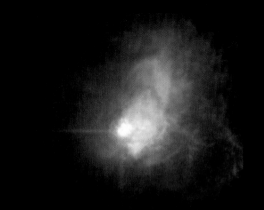

Polarized Light
ACS/HRC

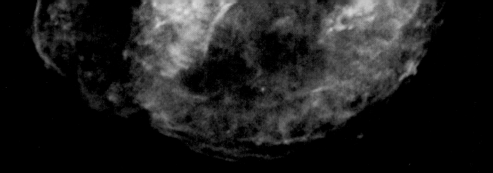

■ *9 January 2007. A spectacular image of Kepler's supernova remnant produced by the Chandra X-ray Observatory helped solve a long-standing puzzle of what kind of supernova produced the debris. In this image, the lowest energy X-rays are shown red, yellow and green are X-rays of intermediate energy, and blue reveals the highest energy X-rays, those emitted from the leading edge shock front of the explosion. The presence of oxygen and iron match amounts seen in Type Ia supernova explosions. These are caused by the thermonuclear explosion of a white dwarf star. Image courtesy NASA/CXC/NCSU/S.Reynolds et al.*

9 January 2007

Seven New Dwarf Galaxies of the Milky Way Found

The number of known dwarf galaxies orbiting the Milky Way has taken a big leap, thanks to discoveries made by a team sifting through data from the Sloan Digital Sky Survey. The new dwarf galaxies contain only a few million stars and have very low surface brightness. The discoveries were made by searching the SDSS database with computer algorithms, searching for similar stars within small regions of the sky.

"We've found almost as many new Milky Way satellites as were detected in the previous 70 years," says Vasily Belokurov of Cambridge University. He notes that they "look as though they are being ground up." The dwarfs, particularly the Ursa Major II dwarf, are being ripped apart by tidal forces as a result of their close passage to the Milky Way. Astronomers believe we are seeing the leftovers from mergers that occurred billions of years ago, early in the formation history of our galaxy.

The team, including Daniel Zucker, also of Cambridge University, made the discoveries in less than one year. More are expected, since the region surveyed represents only one-fifth of the sky. An eighth dwarf, Leo T, lies 1.4 million light-years away, far enough to perhaps be an isolated dwarf galaxy.

The new satellites of the Milky Way come as good news to theorists, who have predicted that there should be tens or hundreds of dwarf galaxies based on Cold Dark Matter models of the universe. Zucker notes, "These discoveries bring the data and the theory closer together."

http://www.sdss.org/news/releases/20070109.dwarfs.html

http://arxiv.org/abs/astro-ph/0701154

http://arxiv.org/abs/astro-ph/0611157

9 January 2007

Kepler's Supernova Remnant: A Star's Death Comes to Life

The location of one of the most famous historic supernovas, Kepler's Star, has come under the scrutiny of the Chandra X-ray observatory, and the dramatic image tells a complex story. The supernova, which exploded before the invention of the telescope over four hundred years ago, has always puzzled astronomers, because the identity of the progenitor remained unsolved.

Size of
Saturn's Orbit

■ 9 January 2007. This representative
color image of the Mira Ceti star
system uses combined data from
the Hubble Space Telescope and the
Keck and Gemini Observatories. Blue
represents Hubble data, and red and
green represents the Keck and Gemini
data. The green color reveals the dusty
outflow from Mira A, while the red
color around Mira B reveals heating
of the outer edge of its disk. Nearly
1 percent of material ejected from
Mira A is captured by the gravity of
Mira B, forming an opaque disk. Image
courtesy Michael Ireland, Caltech.

The absence of a neutron star pointed astronomers, including researcher Stephen Reynolds of North Carolina State University, towards a Type Ia supernova, the violent demise of a white dwarf star. However, the supernova is expanding into a dense region rich in nitrogen, suggesting an explosion of a massive single star that shed a lot of material before exploding.

The X-ray observations may have solved the long-standing puzzle. In the expanding nebulosity known as Kepler's supernova remnant, the relative amounts of oxygen and iron atoms match those seen in Type Ia explosions. In addition, new evidence of "prompt" Type Ia explosions (see 4 January news story) suggests that Kepler's Star may well have been the nearest example of just such a rare event. In these explosions, the progenitor is more massive than usual, and evolves more quickly, reaching its death throes in 100 million years rather than the typical several billion years.

Kepler's supernova remnant will be the focus of more research because of the value of understanding Type Ia explosions and their use as standard candles for cosmological distance measurements.

http://chandra.harvard.edu/photo/2007/kepler/

http://arxiv.org/abs/astro-ph/0703660

http://arxiv.org/abs/astro-ph/0410081

9 January 2007

A New Kind of Planet-Forming Disk in a Famous Star System

Mira Ceti is the prototypical red giant long period variable star in its late stages of life. It is certainly not a place to go hunting for newly forming planets, but once again, nature has proved to be versatile. Mira sheds mass as it goes through its convulsions, flinging its outer layer into space. However, Mira has a companion, Mira B, with an orbital period of about one thousand years. New observations with the 10-meter Keck telescope in Hawaii and the 8-meter Gemini South telescope in Chile reveal some of the ejected material is being attracted to the companion through its gravitational attraction, and is forming a dusty disk capable of forming planets.

"When looking at one of the most celebrated and well-studied stars in the galaxy, I was amazed to find something new and unexpected!" said one of the team members, Michael Ireland of the California Institute of Technology. "The discovery not only changes the way we think about a star that's important historically, but also how we'll look at similar stars in the future."

The radiation from Mira A is heating the edge of the disk, causing it to dimly glow in the infrared. The overwhelming light from Mira A makes seeing the disk very difficult. Using high contrast techniques, this dim glow was observed

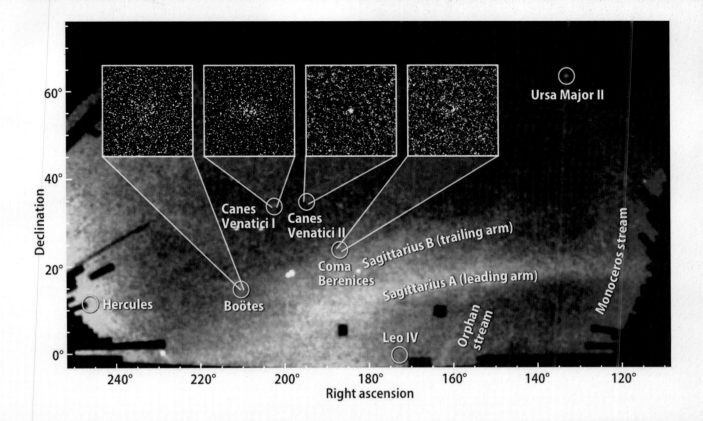

Declination — Right ascension

60° — Ursa Major II

40°

Canes Venatici I — Canes Venatici II

Sagittarius B (trailing arm)

20° — Coma Berenices — Sagittarius A (leading arm)

Hercules — Boötes — Monoceros stream

Leo IV — Orphan stream

0°

240° 220° 200° 180° 160° 140° 120°

Right ascension

for the first time. Mira B is a sun-like star, less massive than our own Sun. Once Mira A ends its life in about a million years and turns into a white dwarf, growth of the new disk will cease, and potentially new planets could form.

It came as a real revelation to see this faint mote of dust, harboring all the possibilities of new worlds in formation, against the hostile environment of the Red Giant," said Peter Tuthill from the University of Sydney, Australia.

http://www.keckobservatory.org/article.php?id=99

http://arxiv.org/abs/astro-ph/0703244

10 January 2007

Chandra Discovers Light Echo from the Milky Way's Black Hole

The Chandra X-ray Observatory has found evidence of a light echo from the most recent large outburst of our galaxy's central black hole. Changes in the shape and brightness of gas clouds nearby Sagittarius A* match those predicted if the black hole swallowed a large mass. Prior to it disappearing forever, a burst of X-radiation would propagate outwards and illuminate clouds of material in the surrounding region.

The changes were noticed between observations made in 2002, 2003, and 2005.

■ **Above:** *9 January 2007. A selection of the eight new dwarf galaxies discovered in the Sloan Digital Sky Survey (SDSS-II) is shown in this plot of the SDSS data. Theoretical models show that many more dwarf galaxies should exist around the Milky Way than have been observed. The new discoveries fill in part of the 'missing dwarf' problem in the Milky Way environment. The small systems of stars were probably cannibalized by passage through the disk of the Milky Way billions of years ago. Image courtesy Vasily Belokurov, SDSS-II Collaboration and reproduced by permission, © 2007 Astronomy magazine, Kalmbach Publishing Co.*

The initial outburst took place 50 years ago when the black hole swallowed a mass equivalent to the mass of Mercury. The reflected X-rays took a longer path to reach Earth and have been detected by Chandra.

The clouds themselves appear filamentary. The observed variability in the clouds spans a number of light-years, suggesting the outburst itself lasted over a couple of years at least. The observations allow astronomers to reject the possibility that the X-ray reflection came from a supernova outburst (they last a short period) or other astrophysical phenomena. The luminosity of the outburst matches that predicted from a large mass being devoured by the central supermassive black hole that lies at the heart of the galaxy.

"Our data show it has been 50 years or so since the black hole had its last decent meal," said Michael Muno of the California Institute of Technology in Pasadena. "This is nothing like the feasting that black holes in other galaxies sometimes enjoy, but it gives unique knowledge about the feeding habits of our closest supermassive black hole."

Chandra has observed more recent outbursts from the central black hole, but they are about a thousand times fainter than the brightness of the original outburst that created the light echoes. If an X-ray telescope had been in orbit 50 years ago, it would certainly have captured the event.

http://chandra.harvard.edu/press/07_releases/press_011007.html

http://arxiv.org/abs/astro-ph/0611651

18 January 2007
Integral X-Ray Observatory Sees the Galactic Centre Playing Hide and Seek

The center of our galaxy is a crowded place, a location of immensely energetic activity – demonstrated by the typically copious X-ray radiation. When the European X-Ray observatory, Integral, pointed toward the region, scientists were understandably surprised and intrigued when they saw a moment of calm. Curiously, ten of the bright sources near the center of the galaxy all faded simultaneously.

"All the sources are variable and it was just by accident or sheer luck that they had turned off during that observation," said Eric Kuulkers of the European Space Agency.

Integral was able to detect much fainter sources previously undetected in its regular survey of the galactic center, called the Integral Galactic Bulge monitoring program, which began in 2005.

"When these normally bright sources are faint, we can look for even fainter sources", said Kuulkers. Most of the eighty or so sources Integral follows are X-ray binaries, but the dimming may allow detection of fainter radiation from the region around the central supermassive black hole at the center of our galaxy, an object called Sagittarius A*.

http://www.esa.int/SPECIALS/Integral/SEMGOVRMTWE_0.html

http://arxiv.org/abs/astro-ph/0603130

http://arxiv.org/abs/astro-ph/0701244

31 January 2007
Hubble and Spitzer Probe Layer-cake Structure of Alien World's Atmosphere

An extrasolar planet known as a "hot Jupiter" is orbiting so close to its parent star that its baked atmosphere is streaming off into space in a huge comet-like stream of gas. The planet, called HD 209458b, orbits every 3.5 days at a distance of 7.6 million kilometers. That is just one-eighth the distance of Mercury from our Sun. The new results come from observations made by the Hubble Space Telescope in 2003, combined with supporting observations using the Spitzer Space Telescope.

HD 209458b lies in Pegasus at a distance of 150 light-years from Earth. It's the first extrasolar planet found to be transiting its parent star from our line of sight. Such transits occur when the planet crosses the disk of the star, providing astronomers with the opportunity to see what happens to the star's light as it passes through the planet's atmosphere. The chemical makeup of the atmosphere is imprinted on the star's spectrum. Such a 'fingerprint' has already revealed oxygen, carbon, and sodium in the planet's atmosphere reported in the initial 2003 results.

Returning to the archived data, Gilda Ballester of the University of Arizona in Tucson, and her team, studied the hot hydrogen atoms in the planet's upper atmosphere, and this led them to the discovery of the leaking atmosphere. The atmosphere becomes bloated by ultraviolet light from the star and is leaking away at a rate of 10,000 tons a second.

"The layer we studied is actually a transition zone where the temperature skyrockets from about 730 degrees Celsius to about 15,000 degrees Celsius, which is hotter than the Sun," said Ballester. "With this detection we see the details of how a planet loses its atmosphere." (See also 2 February news story.)

http://hubblesite.org/newscenter/archive/releases/2007/07/

http://arxiv.org/abs/astro-ph/0401457

■ *31 January 2007. An artist's impression shows the escaping atmosphere of the giant extrasolar planet, HD 209458b. The planet lies a mere 7.6 million kilometers from its parent star, a yellow dwarf. The atmosphere of the planet balloons out, with some gas escaping into space. HD209438b orbits the star every 3.5 days, producing a transit event each time it passes between us and the star. Light filtering through the planet's atmosphere was analyzed by the Hubble Space Telescope. Image courtesy NASA, ESA, and Alfred Vidal-Madjar (Institut d'Astrophysique de Paris, CNRS, France);*

FEBRUARY 2007

2 February 2007

Oxygen and Carbon Found in Extrasolar Planet's Atmosphere

In a further development following the 31 January news from the Hubble Space Telescope, researchers from the Institute d'Astrophysique de Paris in France announced they have found oxygen and carbon in the atmosphere of the transiting planet, HD 209458b. Astronomers were quick to point out that oxygen occurs naturally in gaseous planets, and is not a sign of life on the huge planet.

Dr. Alfred Vidal-Madjar and colleagues report the atmosphere is being ripped off by the "hydrodynamical drag" created by the rapid loss of the previously discovered hydrogen atmosphere. (See 31 January news story for more details.)

http://hubblesite.org/newscenter/archive/releases/2004/44/

http://arxiv.org/abs/astro-ph/0611174

12 February 2007

Comets Clash at Heart of Helix Nebula

The Spitzer Space Telescope has revealed a new appearance to the famous Helix Nebula, and turned up some dust-ridden chaos going on near the central white dwarf. Scientists were surprised by how much dust remained around the dying central star, and they concluded that a large number of comets, thought to exist around all stars, must have survived the original explosion.

The red object at the center of the nebula is the dust-shrouded white dwarf. Such dense objects are formed when a sun-like star, nearing the end of its life, ejects its outer shell of gas, forming the ever expanding nebula. It's a transformation at the end of a star's life that our Sun will pass through in about five billion years time. Astronomers expected to find that any comets orbiting the star, similar for example to those found in our Kuiper Belt, would be ejected by the explosion, leaving a bare, dust-free, white dwarf.

Such a finding is not entirely new, however. A year ago another white dwarf was found with a more compact dust disk extending out to only 0.03 astronomical units (about 4.5 million kilometers).

"Finding evidence for planetary activity around a white dwarf is a surprise," said George Rieke of the University of Arizona, a co-author of the paper. "Finding it twice with such different properties is a shock!"

The disk around the central star of the Helix is much larger than the earlier discovery. It ranges from about 35 AU out to 150 AU from the white dwarf (1 AU is 149.6 million kilometers or about 93 million miles). The total mass is slightly more than one-tenth of an Earth mass scattered throughout the disk.

"The dust must be coming from comets that survived the death of their sun," concludes project team leader Dr. Kate Su of the University of Arizona, Tucson. The explosion must have jostled comets in their orbits, causing a large number of collisions and generating the dust observed by Spitzer. (An image of the Helix Nebula is shown on the front cover of this book. Larger digital versions can be found at the link below.)

http://www.spitzer.caltech.edu/Media/releases/ssc2007-03/index.shtml

http://arxiv.org/abs/astro-ph/0702296

13 February 2007

Colorful Demise of a Sun-Like Star

A beautiful, butterfly-shaped nebula revealing the complex structure of the death of a star is portrayed in a new image from the Wide Field Planetary Camera 2 aboard the Hubble Space Telescope. The planetary nebula, called NGC 2440 (NGC = New General Catalog), is located 4,000 light-years away in the direction of the constellation of Puppis.

A planetary nebula is the result of an explosive release of the outer layers of a dying star whose mass is too small to trigger a supernova explosion. Stars similar in mass to the Sun meet such a fate once their cores run out of hydrogen necessary for fusion reactions. The remaining core of the star is called a white dwarf. The gaseous cocoon resembles the disk of a faint planet in small telescopes, hence their name, except there is no other link other than a morphological similarity to planetary disks.

Hubble's camera has special filters tuned to emission of certain chemical elements. A series of black and white images through each of the filters is used to recreate all color images from the Space Telescope. The resulting digital images from the telescope can be color coded to represent

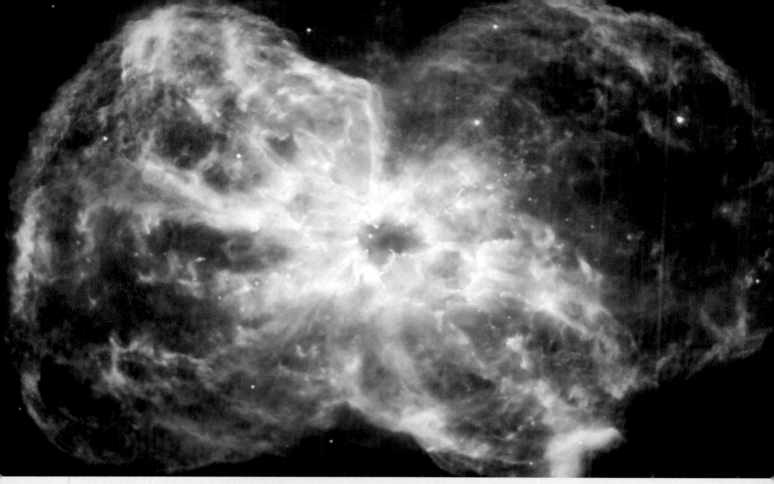

the different elements. In the NCG 2440 image, helium shows as blue, oxygen as blue-green, and nitrogen and hydrogen as red. Ultraviolet light from the white dwarf causes these elements to glow, creating the colorful spectacle.

http://www.spacetelescope.org/news/html/ heic0703.html

13 February 2007

LIGO and Virgo Join Forces in Search for Gravitational Waves

The search for gravitational waves is entering a new era, following a landmark agreement that will enable full sharing of data from two of the newest observatories. The Laser Interferometer Gravitational-Wave Observatory (LIGO) is in the midst of a two-year run at its full operational level and sensitivity. LIGO and its GEO600 partner in Germany is teaming up with the Virgo interferometric gravitational-wave detector of the European Gravitational Observatory (EGO) near Pisa, Italy.

"The combined data will give us a much better chance of finding the first gravitational waves, and will allow us to have greater confidence in any detections," said Benoit Mours of the Laboratoire d'Annecy-le-Vieux de Physique des Particules, representing the Virgo consortium.

Gravitational waves were predicted in Albert Einstein's General Theory of Relativity. Merging black holes and neutron stars are among prime candidates for the first sources to be detected.

LIGO and Virgo have immediate plans for a significant upgrade, increasing their sensitivity by a factor of ten. The improved detectors begin construction in 2008.

http://www.ligo.caltech.edu

http://geo600.aei.mpg.de

http://www.virgo.infn.it

(Note: See also the review article 'The Search for Gravitational Waves … Listening to Space with LIGO' by Laura Cadonati in last year's volume, State of the Universe 2007).

■ **Above:** *13 February 2007. The spectacular and colorful demise of a sun-like star is recorded in this Hubble Space Telescope image captured on 6 February 2007 by the Wide Field Planetary Camera 2. This is a planetary nebula called NGC 2440. Filters used for the image isolate helium, shown as blue, oxygen shown as blue-green, and nitrogen and hydrogen shown as red. At the center of the nebula is a white dwarf, the stellar corpse of a main-sequence star that reached the end of its life. The star may have gone through multiple eruptions as it neared its end. Its surface temperature is a scorching 200,000 degrees Celsius. The nebula lies about 4,000 light-years away. Image courtesy NASA, ESA, and K. Noll (STScI).*

21 February 2007

First Astrophysical Results with the AMBER-VLT-Interferometer

A remarkable new instrument called AMBER (Astronomical Multi-BEam Recombiner) has produced its first results at the Very Large Telescope Interferometer (VLTI) of the European Southern Observatory in Chile.

AMBER is capable of combining infrared light from three of the 8.2-meter telescopes of the VLT, producing extremely high resolution using infrared interferometry. The AMBER/VLTI system can produce resolution 16 times higher than a single 8.2-meter telescope.

The international team that built the instrument was led by Gerd Weigelt of the Max-Planck-Institute for Radio Astronomy in Bonn, Germany. "Thanks to its high spatial and spectral resolution, AMBER/VLTI opens a wealth of completely new insights in various fields of stellar astrophysics", said Weigelt. "This covers the whole range from the very early phases in the life of stars when the star is still accreting surrounding material, to the very late phases when large amounts of matter are blown away from the stars."

Here's a brief look at the new results:

■ *A study of two pre-main sequence stars, MWC 297 and HD 104237, reveals highly detailed data about the outflowing winds from the young stars. The relative aspects of the surrounding gas and orbiting dust disks allow astronomers to gain a deeper understanding about how planets form.*

■ *Eta Carinae, one of the most massive stars known, produced a massive eruption about 160 years ago. New details of the inner region around the* star confirm that ongoing mass loss from the star is not symmetrical. The mass-loss is enhanced along the polar rotation axis. The star itself is shrouded by the dense stellar wind out to about 10 astronomical units.

■ *The double star, Gamma Velorum, contains a Wolf-Rayet star and an O-type star. Both components are hot with strong stellar winds. The AMBER instrument studied the region where these winds collide, and also determined a new distance of 1,200 light-years to the double star, significantly different to the 840 light-year distance determined by the Hipparcos parallax mission.*

■ *The recurrent nova, RS Ophiuchi, was observed five days after its most recent eruption in February 2006, allowing detailed studies of the dynamics of the explosion in its earliest phases. The observations of the expanding fireball revealed a slow expanding ring structure (moving at 1,770 kilometers per second), and a higher velocity east-west structure (moving at 2,400-2,900 kilometers per second). (See also 19 July 2006 news story.)*

http://www.eso.org/outreach/press-rel/pr-2007/pr-06-07.html

http://www.mpifr-bonn.mpg.de/public/pr/amber2007/pr-amber2007-en.html

http://arxiv.org/abs/astro-ph/0606484

http://arxiv.org/abs/astro-ph/0609715

http://arxiv.org/abs/astro-ph/0611602

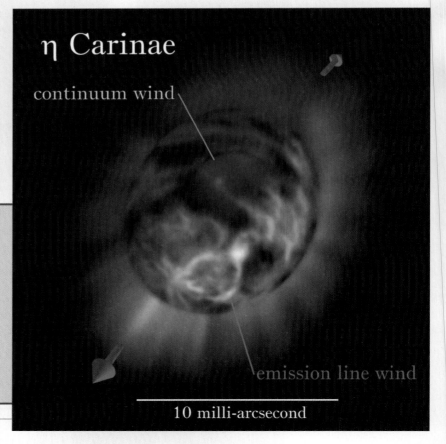

■ *Right: 21 February 2007. An artist's impression of the inner region of the massive star Eta Carinae, interpreted using infrared observations from the new AMBER instrument and three 8.2 meter telescopes of the European Very Large Telescope Interferometer (VLTI). The elongated stellar wind region is shown in blue and displays a continuous spectrum. The larger extended region reveals emission lines of hydrogen, shown in red. This is the first time such detail has been rendered visible. The stellar wind is so dense that it totally obscures the underlying central star. The result is one of a batch of exciting new very high resolution observations made by AMBER/VLTI. Image courtesy Stefan Kraus, Max-Planck-Institut für Radio Astronomie.*

η Carinae

continuum wind

emission line wind

10 milli-arcsecond

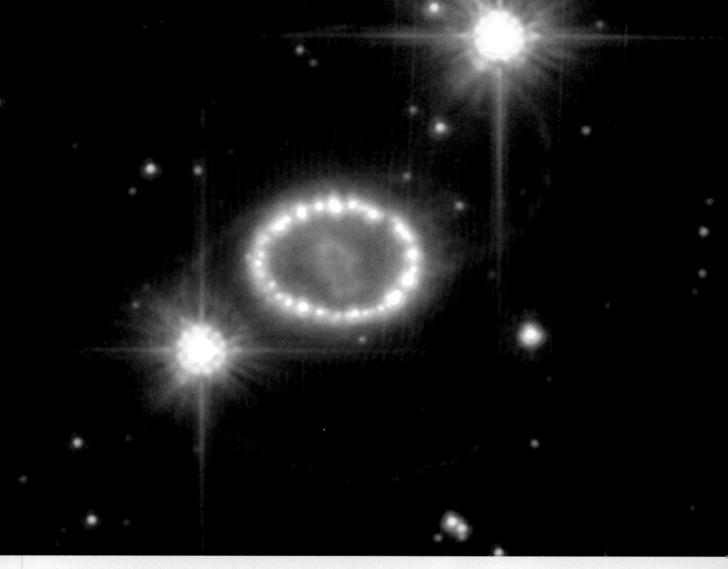

21 February 2007

NASA's Spitzer First to Crack Open Light of Faraway Worlds

While the Hubble Space Telescope has been tracking atomic species in the atmosphere of a distant planet, its sister infrared observatory, the Spitzer Space Telescope (SST), has been following the molecules. One jovian-sized planet in particular, with the fastest known period around its parent star of 2.2 days, is called HD 189733b, and came under SST's scrutiny. It held a surprise for astronomers. They also observed HD 209458b, already known to contain hydrogen, oxygen and carbon from recent HST results.

Astronomers expected to find molecules like water, methane, or carbon dioxide in the infrared spectrum of the two planets. Independent teams of theorists had predicted extrasolar planets would show such strong features. Instead, there were "no molecular footprints" of these molecules.

"In a sense, we're getting our first sniffs of air from an alien world," said David Charbonneau of the Harvard Smithsonian Center for Astrophysics. "And what we found surprised us. Or more accurately, what we *didn't* find surprised us." Charbonneau's colleague, Carl Grillmair of the Spitzer Science Center, added, "We expected to see common molecules like water, methane, or carbon dioxide, but we didn't see any of those."

■ *Above: 22 February 2007. The Hubble Space Telescope reveals the scene of the bright supernova explosion that was seen on 23 February 1987, known as SN1987A. Twenty years later, the expanding nebula has taken on a dramatic 'string of pearls' appearance as the shock wave collides with a ring of gas ejected from the star about 20,000 years before the supernova explosion. The purple region in the center is the expanding debris from the blast, and will continue expanding for decades, eventually colliding with the main ring. The outer pair of rings, faintly visible, remains a mystery. Two bright stars lie in the same field of view but are not associated with the supernova. SN1987A is located 163,000 light-years away in the Large Magellanic Cloud. Image courtesy NASA, ESA, P. Challis and R. Kirshner (Harvard-Smithsonian Center for Astrophysics).*

Both planets appear drier and cloudier than predicted. HD 209458b shows evidence of silicates, suggesting a very dusty atmosphere. Dr. Jeremy Richardson of NASA's Goddard Space Flight Center, Greenbelt, Maryland, thinks this is a major clue for the lack of a signal from methane or water.

"It is virtually impossible for water, in the form of vapor, to be absent from the planet, so it must be hidden, probably by the dusty cloud layer we detected in our spectrum," said Richardson.

Over 200 exoplanets are known to date, and 14 are transiting planets. In addition to the two observed by Spitzer, three more are good candidates for acquiring a spectrum of their atmospheres. Ultimately astronomers would like to have spectra from smaller rocky planets in their quest for the existence of life on other planets.

http://www.spitzer.caltech.edu/Media/releases/ssc2007-04/index.shtml

http://arxiv.org/abs/astro-ph/0702494
http://arxiv.org/abs/astro-ph/0702507

22 February 2007

Supernova 1987A's 20th Anniversary

On 23 February 1987, a brilliant new star blazed in the skies of the southern hemisphere. Viewing a recently developed photograph of the Tarantula Nebula at Las Campanas Observatory in Chile, Canadian astronomer, Ian Shelton, immediately noticed something different about the region in the Large Magellanic Cloud – a new star had appeared. It turned out to be the nearest supernova (SN1987A) since the invention of the telescope.

In the ensuing twenty years since Shelton's discovery, nearly every major telescope that can see the supernova has studied the remnant of the cataclysm. One telescope in particular has revolutionized our view of these stellar explosions. The Hubble Space Telescope has monitored the rapidly expanding debris cloud nearly every year since its eruption.

"The sharp pictures from the Hubble telescope help us ask and answer new questions about Supernova 1987A," said Robert Kirshner, of the Harvard-Smithsonian Center for Astrophysics in Cambridge, Massachusetts. "The Hubble observations have helped us rewrite the textbooks on exploding stars. We found that the actual world is more complicated and interesting than anyone dared to imagine. There are mysterious triple rings of glowing gas and powerful blasts sent out from the explosion that are just having an impact now, 20 years later."

HST was launched three years after the supernova, and its first images captured the immediate attention of astronomers and the public alike. A mysterious set of three rings surrounded the supernova. One glowing ring spanned one light-year, and was emitted by the star 20,000 years before the final explosion. The shock wave has been striking this ring for a few years now, causing dense clumps of material to glow. The expanding debris from the supernova is now one-tenth of a light-year long. In the next few years astronomers expect the inner ring to be bright enough to illuminate the star's immediate surroundings.

The ultimate mystery relates to the fate of the left over core of the supernova. Astronomers predicted that either a neutron star or black hole was produced in the explosion, and expected to see evidence within a few years. Twenty years later, the evidence still eludes us.

http://hubblesite.org/newscenter/archive/releases/2007/10

(Note: See also the review article 'First Views of the Birth of a Supernova Remnant' ... Supernova 1987A' by Richard McCray in last year's volume, State of the Universe 2007).

■ **Right:** *22 February 2007. Schematic diagram identifying the main features of the inner debris ring of SN1987A, shown here in close-up. Main diagram (right) courtesy NASA, ESA and A. Feild (STScI), with inset image (left) courtesy NASA, ESA, P. Challis and R. Kirshner (Harvard-Smithsonian Center for Astrophysics).*

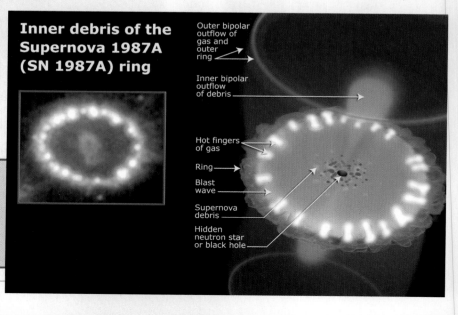

Inner debris of the Supernova 1987A (SN 1987A) ring

Outer bipolar outflow of gas and outer ring

Inner bipolar outflow of debris

Hot fingers of gas

Ring

Blast wave

Supernova debris

Hidden neutron star or black hole

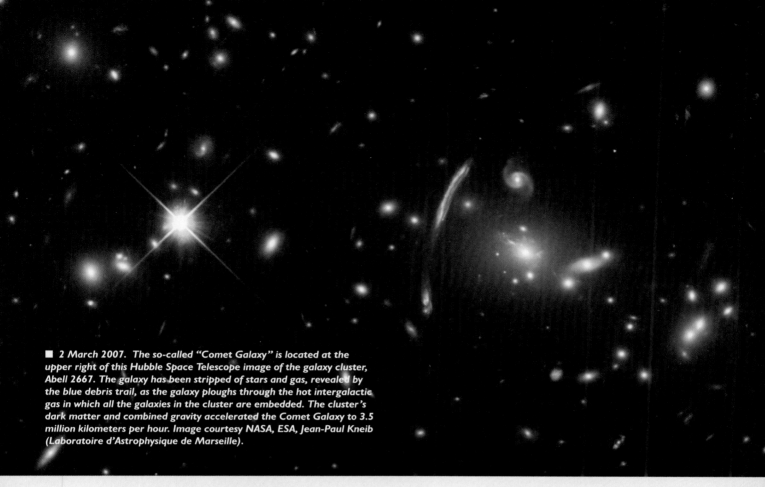

■ *2 March 2007. The so-called "Comet Galaxy" is located at the upper right of this Hubble Space Telescope image of the galaxy cluster, Abell 2667. The galaxy has been stripped of stars and gas, revealed by the blue debris trail, as the galaxy ploughs through the hot intergalactic gas in which all the galaxies in the cluster are embedded. The cluster's dark matter and combined gravity accelerated the Comet Galaxy to 3.5 million kilometers per hour. Image courtesy NASA, ESA, Jean-Paul Kneib (Laboratoire d'Astrophysique de Marseille).*

MARCH 2007

2 March 2007

Hubble Sees "Comet Galaxy" Being Ripped Apart by Galaxy Cluster

Tucked away in the corner of a spectacular image from the Hubble Space Telescope's Advanced Camera for Surveys (ACS) lies the shredded pieces of a galaxy, revealing the aftermath of a comet-like plunge through the outskirts of the giant nearby galaxy cluster. Debris from the collision with intergalactic gas has left pieces of the spiral galaxy in its wake on a much larger scale than anything seen before.

Abell 2667 is a large cluster of galaxies located 3.2 billion light-years away. The ACS image shows the galaxy as it ploughs through the cluster, accelerated to 3.5 million kilometers per hour (2.2 million mph) by dark matter in the huge cluster. As the spiral galaxy collides with the hot intergalactic gas, tidal forces, combined with the pressure from the 10-100 million degree hot gas (ram-pressure stripping), causes an effect on a spiral galaxy similar to a comet ploughing through the solar wind in our solar system. Its appearance led astronomers to call it the "Comet Galaxy". Bright knots from hot blue stars stripped from the parent galaxy lie in its wake.

An international team of astronomers are studying how the gas is stripped from spiral galaxies, and the "Comet galaxy" is an excellent example of one of the mechanisms. They are trying to understand why distant galaxy clusters show only 20 percent of galaxies as gas poor when the universe was roughly 7-8 billion years old, compared to 50 percent today.

"By combining Hubble observations with various ground- and space-based telescopes, we have been able to shed some light on the evolutionary history of galaxies," said Luca Cortese of Cardiff University in Wales, U.K.

Other major telescopes provided a wealth of information about the speeding galaxy, such as the age of its star forming region (ESO's Very Large Telescope and the twin Keck Telescopes), and the origin of the activity as being due to star formation and not a supermassive black hole (the Spitzer Space Telescope and the Chandra X-ray Observatory).

http://www.spacetelescope.org/news/html/heic0705.html

http://arxiv.org/abs/astro-ph/0703012

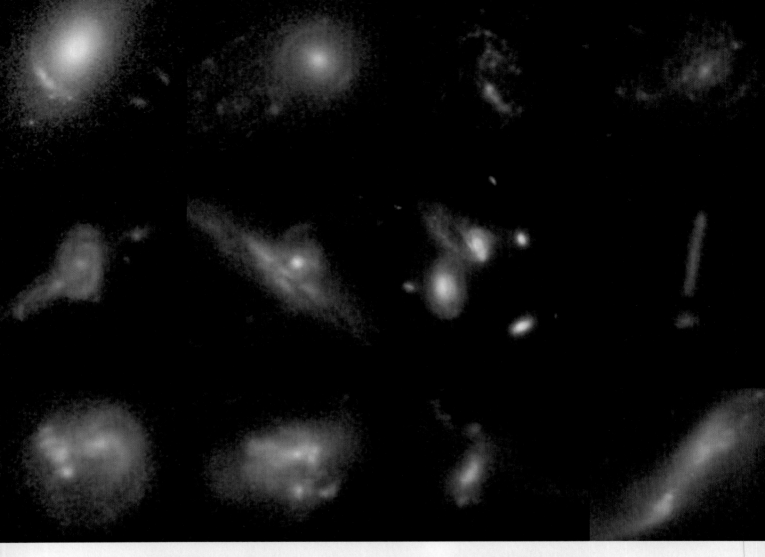

6 March 2007

AEGIS Survey Reveals New Principle Governing Galaxy Formation and Evolution

New results from a massive survey of a small region of the sky using the world's most powerful telescopes is yielding new details of the early youth and evolution of galaxies and of galaxy clusters. The area of sky under scrutiny is called the Groth strip, a 1.1° x 0.15° section of sky in Ursa Major named after Princeton University

■ **Above:** *6 March 2007. An array of unusual galaxies from the AEGIS survey is displayed in this montage. In the top row, two red elliptical galaxies act as gravitational lenses, producing the blue rings called Einstein Rings. The foreground galaxies are distorting light from much more distant galaxies located behind the relatively nearby elliptical galaxies. The middle row shows pairs of galaxies that appear to be in the process of merging, although some of the pairs may be a line of sight effect. The lower row shows irregular galaxies containing bright blue clumps of active star formation. Note the classic "tidal tail" in the image second from right, produced by the gravitational interaction between two merging galaxies. Mergers are predicted to have been more frequent in the early universe. Image courtesy AEGIS/UCOLICK.*

physicist, Edward Groth. Like many regions of the sky, no stars are visible to the unaided eye, but to the Hubble Space Telescope and Keck telescopes over 150,000 galaxies have been found there.

The new observations are part of a five-year study called the All-wavelength Extended Groth Strip International Survey (AEGIS). As the name implies, observations span the entire electromagnetic spectrum, from x-rays to radio.

"The goal was to study the universe as it was when it was about half as old as it is at present," said team leader Marc Davis, professor of physics and astronomy at the University of California, Berkeley. "We've gotten such fabulous data; it just blows your socks off."

The observations have shown a vast array of spiral and elliptical galaxies, as well as amorphous masses that appear to be the remnants of violent mergers of young galaxies.

"We're studying a key epoch when galaxies appear to be taking on their final mature forms," said Sandra Faber of the University of California

at Santa Cruz. "It's like seeing people at the age of 10 – they are not exactly infants, but they differ substantially from adults. We are watching galaxies grow up."

A key component in the discoveries is the DEEP2 Redshift survey which has measured the light from over 14,000 galaxies in the Groth strip using the 10-meter telescopes of the Keck Observatory in Hawaii. The completed survey will include 18,000 galaxies studied by the DEEP2 survey.

Among the major results is the discovery of two supermassive black holes in the core of a giant red galaxy. Two black holes are separated by 4,000 light-years in the core of an active galaxy.

Of deeper significance is the new link between the Tully-Fisher relationship and the Faber-Jackson relationship. The T-F relationship recognizes that the stars and gas in more massive spiral galaxies rotate faster than less massive ones. The F-J relationship is similar, but applies to elliptical galaxies. It links the random motions of stars with the mass of giant elliptical galaxies.

The AEGIS survey discovered a new speed indicator that takes into account both the random and rotational components of velocity, and links it to the mass of the galaxy, irrespective of whether it is elliptical, spiral, or even irregular. This new relationship appears to have held fast for at least the past 8 billion years.

"Surprisingly, if you use this new speed indicator to measure the motions of stars and gas in a galaxy, you can predict the mass in stars the galaxy has with pretty high accuracy," said Susan Kassin, a post-doctoral researcher at the University of California, Santa Cruz. Kasin and colleagues found the relationship holds "for all kinds of odd-ball galaxies that are more common in the early universe".

In addition to Hubble and Keck, other telescopes contributing to these results included Chandra (X-ray), GALEX (ultraviolet) and Spitzer (infrared) orbiting telescopes, and the ground-based Canada-France-Hawaii Telescope, the 5-meter (200-inch) Palomar Telescope, and the Very Large Array (radio). The full set of data should be complete in August 2007.

http://aegis.ucolick.org.

http://aegis.ucolick.org/publications. html#apjl2006

http://arxiv.org/abs/astro-ph/0607355

http://arxiv.org/abs/astro-ph/0608380

http://arxiv.org/abs/astro-ph/0702643

http://arxiv.org/abs/astro-ph/0701924

7 March 2007

ALMA Achieves Major Milestone with Antenna-Link Success

The most advanced telescope for studying millimeter and sub-millimeter wavelengths, the ALMA array, has passed another major milestone in its construction. Two of the prototype antennas have been linked together for the first time and successfully observed the planet Saturn. Acting as a single instrument, the two antennas and data-processing pipeline were fully tested end to end.

■ *Above: 6 March 2007. A collection of strong gravitational lenses has been found in the HST/ACS images of the AEGIS survey. Light from distant galaxies is distorted by intervening galaxies producing highly magnified arcs, multiple images, and in some cases full "Einstein Rings". Here we see a near double-ring (left), a skewed, paired-arc set of images (center), and a classic "Einstein Cross" (right). The AEGIS survey has produced a three-dimensional map of the large scale structure along the lines of sight to these lenses. The total influence of this matter ought to match up with what the gravitational lens models imply. Image courtesy AEGIS/UCOLICK.*

ALMA tracked the 104 gigahertz emissions from Saturn for over an hour. The ALMA array will eventually contain 66 antennas when it is completed in 2012. It is being constructed in the Atacama Desert in northern Chile.

"This achievement results from the integration of many state-of-the-art components from Europe and North America, and bodes well for the success of ALMA in Chile," said Catherine Cesarsky, ESO's Director General.

http://www.nrao.edu/pr/2007/almafringes

8 March 2007

Gamma-Ray Burst Challenges Theory

An unusually long afterglow from a gamma-ray burst detected in July 2006 is resulting in a new look at the theories of what causes certain types of bursts. The Swift Gamma-ray Observatory observed the X-ray afterglow from GRB 060729, located in the constellation of Pictor, for 125 days, far longer than any previous burst on record. Normally they are observed for a week or so at most.

Such a long afterglow requires a different mechanism than the one typically suggested for long bursts, that of a hypernova, the supernova of a very massive star. The slow fading requires a massive injection of energy, perhaps over a longer period than has normally been surmised.

The favored suggestion is that the source was a magnetar, a neutron star with a huge magnetic field. The magnetic field slows the neutron star's rotation. The energy of the spin-down can be converted into magnetic energy, fueling the initial blast wave. Calculations by Xiang-Yu Wang of Pennsylvania State University show this energy is sufficient to keep the afterglow shining for months.

Eleonora Troja, a member of the Italian astrophysical institute, INAF-ISAF, suggested the magnetar origin. "From the properties of the X-ray afterglow, we estimate that the central object should have been rotating with an initial frequency of about 1,000 times per second," says Troja. "That initial frequency is very close to the break-up speed of a spinning neutron star."

"People have thought for a long time that GRBs are black holes being born, but scientists are now thinking of other possibilities," says Swift principal investigator Neil Gehrels of Goddard Space Flight Center.

http://www.science.psu.edu/alert/Grupe3-2007.htm

http://arxiv.org/abs/astro-ph/0611240

12 March 2007

LBT Captures Extremely Faint Light with its First Mirror and Camera

With the detection of the faint afterglow from a gamma-ray burst that occurred almost a month prior, the brand new Large Binocular Telescope has proved its worth. The 26th-magnitude glow was detected during a "science demonstration" program designed to show astronomers that the twin 8.4-meter telescopes will deliver the science it was designed to perform. First light occurred in October 2005, and the first observations with the twin beams from both mirrors will occur in late 2007. The twin telescope is designed to have the resolution of a 22.8 meter telescope.

http://medusa.as.arizona.edu/lbto/

22 March 2007

Gemini's Laser Vision Reveals Striking New Details in Orion Nebula

The new ALTAIR laser guide star adaptive optics system on the Gemini telescope has provided astounding detail of the outer regions of the Orion Nebula. The new image reveals "bullets" of gas driving through the nebula, leaving in their wake a regular series of waves in the molecular hydrogen gas of the Orion Nebula.

The fine details are made possible by a deformable mirror that counteracts the turbulence of the atmosphere as revealed by an artificial star generated by a laser beam shining on a layer of our atmosphere. The fast corrections of the deformable mirror allow the distortions to be removed in real time.

The large clumps of gas, many times the size of our solar system, are speeding out of the nebula at 400 kilometers per second. They've been recognized since 1983, but have never been seen in such detail before. Their origin remains a mystery, although scientists think they are less than a thousand years old.

The high level of detail excites astronomers like Michael Burton of the University of New South Wales, Australia, According Burton, "This level of precision will allow the evolution of the system to be followed over the next few years, for small changes in the structures are expected from year to year as the bullets continue their outward motion."

http://www.gemini.edu/bullets

■ **Above:** *22 March 2007. A small region of the Orion Nebula reveals interstellar "bullets" of material traveling at 400 kilometers per second (250 miles per second) through the surrounding nebula. The image was taken using the ALTAIR adaptive optics system on the 8-meter Gemini North telescope, located on Mauna Kea, Hawaii. Carefully selected narrow-band filters isolate the light from hot iron gas that causes the tip of each bullet to glow blue, while each bullet leaves a tubular pillar that glows orange by the light of excited hydrogen gas. Image courtesy Gemini Observatory/AURA/NSF.*

26 March 007

Initial Results from the AKARI Infrared Satellite

A flood of new results from the Japanese orbiting AKARI infrared observatory is shedding new light on star formation, star death, and the evolution of galaxies. AKARI is performing a sensitive all-sky survey in the infrared using its 68.5 cm telescope, spanning wavelengths from 1.7 to 180 microns. The new results were presented at the Astronomical Society of Japan.

Some major highlights were:

■ *At least three generations of continuous star formation have been found in the same nebula, with the new generation depending on activity of the previous generation. Parent stars trigger pressure waves in the interstellar medium, leading to the following generation of stars.*

■ *A new infrared detection of a supernova in the neighboring satellite galaxy, the Small Magellanic Cloud, reveals details of how the expanding shell interacts with the surrounding interstellar medium.*

■ *Red giant stars are theoretically known to loose mass in the early stages of their evolution. While observations have been made of this process in older red giant stars, AKARI has made the first observations of young red giants, providing much needed data.*

■ *AKARI has made the first detection of carbon monoxide in the vicinity a black hole at the center of an active galaxy. Such black holes are often hidden from view by optically thick dust surrounding the active core, but infrared light penetrates such dust.*

■ *New evidence from a deep infrared survey by AKARI has shown that the intense period of star formation thought to have occurred about 6 billion years ago began, in fact, much earlier.*

http://www.ir.isas.jaxa.jp/ASTRO-F/Outreach/index_e.html
http://arxiv.org/abs/0704.0706
http://arxiv.org/abs/astro-ph/0605589

■ *Opposite page: 26 March 2007. A color composite of the IRC4954/4955 nebula taken by the AKARI infrared satellite. White dots are infrared stars. A dark hollow region in the center of the image has a diameter of about 100 light-years.*
Image copyright © 2007 JAXA/ISAS/LIRA.

28 March 2007

VLT Automatically Takes Detailed Spectra of Gamma-Ray Burst Afterglows Only Minutes After Discovery

The exceedingly faint optical afterglow of a Gamma-Ray Burst is hard to detect, and to make matters more difficult, it can fade by a factor of 500 in the first few minutes following an outburst. Observing the spectroscopic details of the afterglows is a kind of "holy grail" for astronomers working at unraveling the mysteries of gamma-ray bursts. Consequently the best chance of capturing a spectrum is with a large telescope that can react very quickly, a technical challenge of engineering and quick response.

The European Southern Observatory's Very Large Telescope (VLT) is equipped with a Rapid Response Mode, allowing robotic activation of an observing sequence the moment an alarm from the Swift Gamma-ray Observatory is sounded. Success came on 17 April 2006 when the VLT acquired a spectrum less than ten minutes after the Swift observatory spotted the GRB. This was followed on 7 June by a spectrum obtained only 7.5 minutes after the burst.

The high resolution spectra were obtained by the Ultraviolet and Visual Echelle Spectrograph (UVES), mounted on Kueyen, the second Unit Telescope of the Very Large Telescope.

"The afterglow spectra provide a wealth of information about the composition of the interstellar medium of the galaxy in which the star exploded. Some of us even hoped to characterize the gas in the vicinity of the explosion," said team member Cédric Ledoux (ESO).

A cloud of material set aglow by the burst, known as GRB 060418, was found to be located 5,500 light-years from the origin of the burst, much farther than anticipated, and is the first ever such measurement from a GRB. The commonly-held view that afterglows come from regions close to the GRB may have to be revised.

http://www.eso.org/outreach/press-rel/pr-2007/pr-17-07.html
http://arxiv.org/abs/astro-ph/0611690

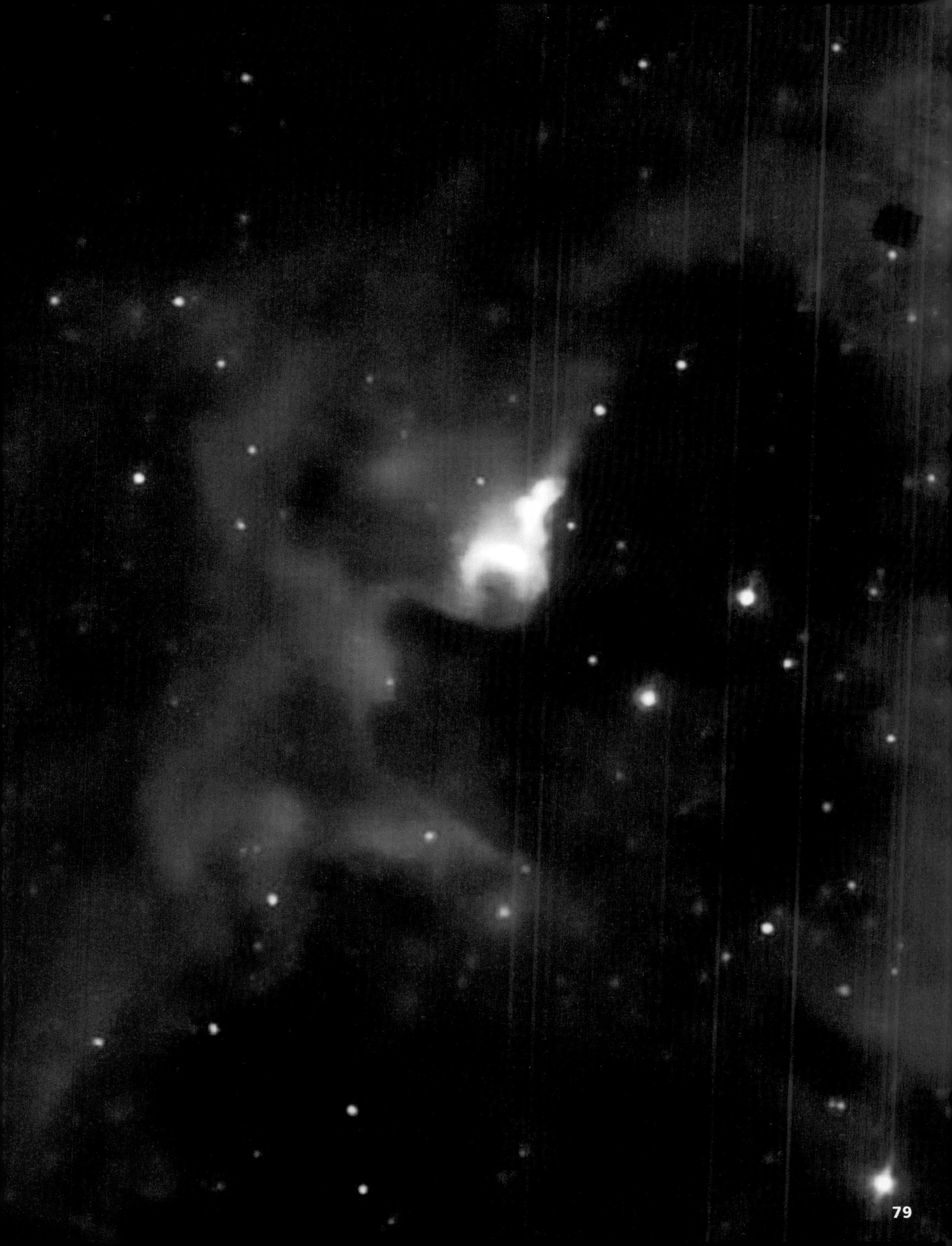

2

Astronomers Gather
IN SEATTLE...

In January 2007, over 2,200 astronomers, physics teachers, and journalists attended the AAS/AAPT Joint 2007 Winter Meeting in Seattle. *Martin Ratcliffe* was there, interviewing undergraduates, graduates, and professional astronomers, including the President of the AAS, to catch up on what the astronomical research community's been up to, and get some feedback on a dynamic meeting.

...The 209th AMERICAN Astronomical Society MEETING

I T'S AMAZING!" "I've found it invaluable." "Inspirational." These are some of the comments from first timers and seasoned delegates at the 209th meeting of the American Astronomical Society (AAS), held from 5-10 January 2007 at the Washington State Convention and Trade Center in Seattle, Washington. The biennial gathering was held jointly with the Association of Physics Teachers (AAPT), bringing over 2,200 delegates from around the country.

From particle physics to flying on the space shuttle, from the enigmatic gamma-ray bursts to 'hot Jupiters', and from NASA space policy to the newest observations from the farthest reaches of the universe, every lecture hall, discussion panel, and impromptu discussion in coffee bars and restaurants, was filled with the excitement of the newest discoveries, the latest research, as well as the best advice for furthering your career in astronomy.

"It's an excellent way for people doing astronomy all over the world to see the latest projects going on, and to learn about the newest findings in astronomy," said Stephanie Fiorenza, referring to her attendance at the meeting. "I think it's really a great way to see what's going on," she added. Fiorenza, a Penn State undergraduate attending an AAS meeting for the first time, co-authored a poster paper with her supervisor about the gamma-ray burst GRB 041223.

A poster at a professional research meeting is a display of the newest research, complete with graphs, and sometimes images. Each day at an AAS meeting nearly 400 different posters are presented. They give astronomers an opportunity to hear comments, both critical and supportive,

■ *Left: Nathan Smith (UC Berkeley) discusses the observations of rings of material around Luminous Blue Variables, whose morphology resembles rings observed around supernova SN1987A, with Sky & Telescope editor Robert Naeye, and Phil Plait, author of Bad Astronomy and its associated well-known web-site. Image courtesy Martin Ratcliffe.*

the people from all the different telescopes and collaborators."

Kevin Croxall (Indiana State University), attending his fourth meeting since 2000, noticed "in that short space of a few years ... a real difference in the numbers of people attending the meeting," and "now seeing bigger presentations and color posters." Croxall and colleague, Walter Trentadue, presented new observations from the WIYN telescope.

of their research, and a chance to improve their text before it goes for final publication in refereed journals.

The AAS meeting represents a rare opportunity for an undergraduate to rub shoulders with graduates, observatory directors, and seasoned project scientists, who all stroll past each poster paper on display during the four day meeting. Undergraduates represent the newest generation of scientists coming up through the ranks, and they experience their first taste of an active research meeting.

Neighboring Fiorenza's poster was Adria Updike of Clemson University, who is using GRBs to probe the early universe. Updike, a graduate student, is attending her fourth conference. "I've just started work on GRBs for my thesis .... it's great to get out and meet all

■ **Top:** *Modern-day scientists discuss new research papers on display in the large exhibit hall at the Washington State Convention Center. Hundreds of poster papers are presented each day at an AAS meeting. Many of the latest colorful images from space-based and ground-based telescopes illustrate the latest findings. Image courtesy Martin Ratcliffe.*

■ **Above right:** *Walter Trentadue (Indiana University), Amandeep Gill (Brown University), and Kevin Croxall (Indiana University) exhibit their research on lithium abundance measurements in star clusters using the 3.5-meter WIYN consortium telescope (Wisconsin, Indiana, Yale, and National Optical Astronomy Observatory (NOAO)). Image courtesy Martin Ratcliffe.*

Representing the culmination of his research, Trentadue and colleagues are studying how lithium is depleted in the stars of an old open cluster, NGC 188. Trentadue, like many people at the conference, is reaching a critical point in his career as a new astronomer. "So far, (I've met) a lot of very interesting people here, and I'm doing my best to network and make as many contacts as I can, not only for future REU (Research Experience for Undergraduates – funded by the National Science Foundation) progress, but for places to go for graduate school in astronomy," said Trentadue.

He, like many others, is excited about the future, as expressed in his parting words, "I've learned that there is so much that I don't know; there are so many different topics within astronomy. It's amazing, (there's) more than a lifetime's worth of research ahead for me."

These stories are echoed by hundreds of undergraduates and graduates at the conference. Each one is starting out on a new career in

research, making new contacts, and realizing the enormous breadth and depth of the subject. All have different reasons for pursuing astronomy as a career, but share a common purpose among all astronomers worldwide. It's an insatiable desire to understand the universe around us.

The individual personalities that make life interesting, and day-to-day practicalities of living on this tiny blue planet, are unified in the ultimate search for our origins. Every generation has uncovered more of the clues scattered throughout the universe, from the origin of the chemical elements that form everything we have ever known, to the birth and death of stars, and outwards to the expanding, and now apparently accelerating, universe. The puzzle is huge and complex, but every generation adds their own unique piece to the picture that future generations will use to take the next steps in our understanding.

At the upper end of the career spectrum, Professor J. Craig Wheeler, the incumbent

■ **Left:** Two spectacular images of the Kepler Supernova remnant (from the Spitzer Space Telescope), and the Eagle Nebula (Chandra X-ray Observatory) were released to the public, as well as new evidence of ring nebulae around Luminous Blue Variables, and their implications for the progenitor of SN1987A, by astronomers at a press conference during the AAS meeting in Seattle. Astronomers Nicolas Flagey (Institut d'Astrophysique Spatiale, France), Nathan Smith (UC Berkeley), and Stephen Reynolds (North Carolina State University) presented the new findings. AAS photo by Kelley Knight, © 2007 American Astronomical Society.

ceremonial things and chairing many of the plenary sessions," he added.

When asked what news from the meeting struck him the most, Wheeler immediately described the excellent standard and high quality of the plenary sessions given by invited speakers. "I have learned science from them." he said, a reference to the cutting edge nature of the invited talks. Astronomers are always learning new

President of the AAS at the conference, exemplifies the qualities of a professional astronomer. This theoretical astronomer is wide in experience, has the ability to communicate to a broad audience, and has a friendly, approachable nature. He took a few moments from a busy schedule of meetings and chairing sessions to discuss the conference. "I haven't had much time to attend many sessions myself," he confessed. "They keep the President so busy running around doing

things, whatever their seniority – the excitement of continual discovery keeps scientists hooked on astronomy.

Wheeler, like many experienced astronomers at the meeting, has had a lifelong quest for origins. His career spans over five decades, and his research covers topics like supernovae, black holes, and gamma-ray bursts. His stint as AAS President has been at a critical time for astronomers. NASA's budget is undergoing great changes due to internal reorganization for the return to the Moon and the commitment to finish the International Space Station. There's a great strain on the basic science part of the NASA budget, and many careers ride on its funding. Sudden budget changes bring tension, and Wheeler has excelled at developing a professional and intelligent discussion by the society.

In describing the distinguished committee he appointed – the Committee on Astronomy and Public Policy (CAPP) – Wheeler noted that the committee members are "some of the most top level, influential, (and) thoughtful people". At a time when NASA's Advisory structure

■ **Top:** *AAS President, J. Craig Wheeler, shares humor with the 2007 Warner Prize for Astronomy winner, Re'em Sari of Caltech. Sari presented his prize lecture on the formation of the solar system, and the origin of planetary spins. AAS photo by Kelley Knight, © 2007 American Astronomical Society.*

■ **Above right:** *The Committee on Astronomy and Public Policy (CAPP) held a very well attended panel discussion about the future of astronomy and astrophysics at NASA, chaired by Jack Burns (University of Colorado). The panel, shown here, are (from left to right), Neil DeGrasse Tyson, Anneila Sargent, Garth Illingworth, Harrison (Jack) Schmidt, Jack Burns, and Lennard Fiske. Tyson and Fiske are members of the NASA Advisory Council, which is chaired by Harrison Schmidt. Burns, Sargent, and Illingworth represent CAPP. AAS photo by Kelley Knight, © 2007 American Astronomical Society.*

■ **Opposite page top:** *A big story was the first determination of the three-dimensional structure of dark matter, achieved with observations from the COSMOS survey made using the Hubble Space Telescope. The four panelists presenting the news to the science media, during a press conference at the AAS meeting in Seattle, are (left to right); Richard Massey, Nicholas Scoville, Richard Ellis (Caltech), and Jason Rhodes (JPL). Mapping was performed using a weak-lensing technique that produces the lensing effects at different redshifts, enabling the 3-D structure to be determined. (See also 7 January news story in Chapter 1 of this book.) Image courtesy Martin Ratcliffe.*

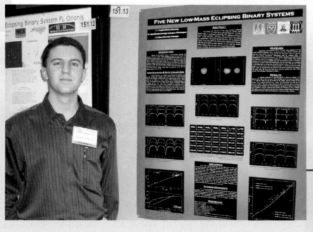

is undergoing change and budgets are being redirected for the return to the Moon, basic astronomy and astrophysics research is under increasing pressure.

Three members of the NASA Advisory Council (NAC), Chair and Apollo 17 lunar geologist, Harrison (Jack) Schmidt, Neil DeGrasse Tyson, and Lennard Fiske joined three members of the CAPP for an open forum on the pressing issues of reduced funding.

Tyson expressed many people's feelings when he commented on the high level of public support that NASA enjoys, using evidence from the Rose Center and Hayden Planetarium in New York City that he directs. He cautioned that we should not "take for granted what has created

■ *Above left: Jeff Coughlin (Emory University), attending his first professional AAS meeting, presents research describing five new low-mass eclipsing binary systems. Image courtesy Martin Ratcliffe.*

■ *Above: L. Jeremy Richardson (Goddard Space Flight Center) discusses his research into studying the atmospheres of extrasolar planets using a clever infrared technique in his poster paper at the Seattle AAS meeting. Because an extrasolar planet is brighter than a star at certain IR wavelengths, sequential spectroscopic observations in the infrared during times with the planet in front of and behind the star allow for subtraction of the star's light, leaving the infrared signature of the planet. Richardson is now the first John Bahcall Public Policy Fellow of the American Astronomical Society, a position established to expose talented research astronomers interested in public policy to the scientific advocacy process. Image courtesy Martin Ratcliffe.*

service HST (a decision reversed in 2006) underline astronomy's significance.

The session talked constructively in, as Wheeler put it, "a non-NASA-bashing way, about what we perceive the problems are" in the current funding battles. As the leader of the astronomical community, Wheeler was clearly pleased with the "sophisticated, intelligent, and thoughtful way" the discussions were handled.

Wheeler enjoys getting involved in the latest work on the dark energy question, in which he is "somewhat engaged". "How do we measure the accelerating universe more precisely?" Surely the most challenging topic in cosmology, Wheeler noted the task of measuring the accelerating universe involves "a lot of hard work to beat down the error bars" by the many teams involved, and expects to see more definitive results "in a year or two from now".

The changing nature of topics at the meeting caught the attention of many delegates. For example, the discovery of the first extrasolar planets just over a decade ago has blossomed into a rich and productive research field. "You wouldn't have found a poster on it in previous years. Now there are oral sessions and dozens

that support," emphasizing that it is "culturally important what science does." The ubiquitous nature of Hubble Space Telescope images, and the public outcry about the decision not to

■ **Top:** *Timothy Hankins (New Mexico Tech) answers questions from Francis Reddy (Astronomy Magazine) and George Musser (Scientific American) during a press conference at the Seattle AAS meeting. Hankins presented new high resolution observations of pulses coming from the Crab Nebula using the 1,000-foot Arecibo radio telescope. Image courtesy Martin Ratcliffe.*

■ **Above:** *James Graham and Paul Kalas (UC Berkeley) announce the discovery of fluffy grains, the first direct indication of the growth of larger dust grains inside a proto-planetary disk, during a press conference at the Seattle AAS meeting. Graham and Kalas discuss the broad topic of proto-planetary disks in their feature article in Chapter 4 of this book. Image courtesy Martin Ratcliffe.*

of posters," said Jeff Coughlin of Emory University, attending his first meeting. "It's very organized with a huge array of topics," Coughlin said, adding it was nice to have his own specialized part of the meeting. Coughlin and colleagues discovered five new low mass eclipsing binary systems.

Jeremy Richardson, (Goddard Space Flight Center, GSFC) is working on cutting-edge research, detecting the atmospheres of extrasolar planets using the fact that some planets are occulted by their parent star. Richardson's paper on infrared spectroscopy of an extrasolar planet (HD 209458) was simply impossible a few short years ago. "It's been a pretty exhilarating year, in particular for the extrasolar planet research that we're studying, so we are very excited," Richardson said.

When the planet is occulted by the star, using the subtraction effect in the spectrum, astronomers can differentiate the planet from the star in the infrared (IR), because, as Richardson's colleague, Craig Jemmings (GSFC) notes, "in IR the planet is a significant source because it's a hot Jupiter."

"One of the big things is that we can directly measure the detected planet now. We can't give you a picture, but we can count the photons," said Jemmings, clearly thrilled with this new and productive line of research.

The extrasolar planet poster was one of many filling the conference display hall. "There are several posters down this line that are pretty exciting, including one talking about a planet around a white dwarf that no one has ever seen before," Richardson added.

Jemmings is a veteran of AAS meetings, attending since the 1970s, and Richardson has attended "quite a few". Reflecting on the excitement at the Seattle meeting, Richardson recounted, "I've just had a grad student come up to me who issued a press release a few days ago about the phase variations of an extrasolar planet …. He's just finished his course work, and is getting ready to start his research, and I said, that's a good way to start your career in astronomy, with a big press release."

Richardson has an additional interest in the broader issues facing astronomy. He is spending six months as the John Bahcall Public Policy Fellow at the headquarters of the American Astronomical Society in Washington D.C. While in the post, he is exploring congressional budget issues, federal science policy, funding levels, and adding to the intelligent discourse about funding espoused by AAS President, J. Craig Wheeler. He also understands the broader impact of astronomy, adding "Astronomy is interesting because it's in a unique position to get young people to think about doing science and math."

Triggering a growing excitement for the future of space-based astronomy was the awesome presence of a full scale model of the James Webb Space Telescope in a courtyard outside the registration area. This huge telescope will open up the infrared universe, when launched in 2013, and will follow in the footsteps of the European Space Agency's 3-meter Herschel infrared telescope due for launch in 2008. JWST is double the size of Herschel, carrying a huge 6-meter diameter segmented mirror, and will peer deeper and farther into space than any telescope before it. Major discoveries and surprises are practically guaranteed. The individual mirror segments are nearly complete. An AAS Town Hall Meeting devoted an entire evening to updates about the

■ **Above left:** *John Mather, recipient of the 2006 Nobel Prize for Physics for his work on the COBE mission, attended the AAS meeting to discuss progress with the James Webb Space Telescope project. Mather is the JWST Project Scientist. AAS photo by Kelley Knight.* © 2007 American Astronomical Society.

telescope's construction and the expected science. Nobel laureate and JWST Project Scientist, John Mather, as well as the lead engineers involved in the project, gave detailed summaries.

Like many national astronomy meetings around the world, AAS conferences create a period of time to measure the current state of our exploration of the universe. The universe is as it is, but our knowledge of its detailed workings improves every year. New results from the latest telescopes mark success of projects that used to be simply plans on paper – often the subject of forward-looking posters at some past AAS meeting. These results sit side-by-side with the newest technological dreams and wishes for the next decade of research. Projects like the Large Synoptic Survey (LSST) and the Thirty Meter Telescope (TMT) are pushing the limits of optical telescope technology. High-energy space-based observatories like GLAST are discussed in detail since launch is near, while others, such as the Supernova/Acceleration Probe (SNAP) and Constellation-X, are in the early proposal stages.

As graduates and undergraduates discover, an AAS meeting is a place to find out what area of research you wish to be a part of, and to keep up with the breathtaking pace of new discoveries. Each new year brings its string of new discoveries. Experienced astronomers gather in team meetings in coffee houses and restaurants discussing their latest observations, and planning the next round, results that should appear at a forthcoming AAS meeting. As everyone discovers by the first day of the meeting, keeping up-to-date is like sipping water from a fire hose – no one person can possibly take it all in.

■ **Above:** *Two of the many display stands at the 2007 AAS meeting in Seattle were those of the National Radio Astronomy Observatories and the Space Telescope Science Institute, providing brochures and posters highlighting recent achievements. Nearly all major observatories and current space-based observatories display at the AAS meetings, as well as many book publishers and major telescope manufacturers. Image courtesy Martin Ratcliffe.*

■ **Opposite page:** *A full-scale model of the James Webb Space Telescope attracted a lot of attention at the 2007 AAS meeting in Seattle. The hexagonal mirror segments sit above large wings that act as insulators to protect the telescope from the heat of the Sun. Control thrusters can be seen to the right of the author, who is shown for scale. JWST was the subject of a "town meeting" reviewing many aspects of the mission. Launch is expected in 2013. Image courtesy Martin Ratcliffe.*

3

The State of
THE UNIVERSE...

In their hunt for Dark Matter, Bunny thought the Human Astronomers could be an unobservant lot!!

Returning for a second year, *Jim Kaler* offers a personal perspective on the new slate of significant discoveries in 2006-07, ranging from individual stars to the origin of the universe itself.

...New Steps in
EXPANDING
our
KNOWLEDGE

IT'S IN fine shape. Flat. That is, it conforms to Euclid's ancient laws of geometry, as modified 2200 years later for Einstein's relativity and spacetime. At its expansion rate of 73 kilometers per second per megaparsec (a megaparsec equals 32.6 million light-years), over the past year the Universe and its expanding spacetime has increased its 'size' by 0.0000000073 percent. Of course we don't really <u>know</u> its 'size' – the term really refers to the separation scale among the things we can see.

The 'Big Event' this year was the measurement of the polarization (alignment) of radio waves from the Cosmic Microwave Background (CMB) as viewed by the WMAP satellite, which with additional information gives us the expansion rate, an age since the Big Bang of 13.7 billion years, a 'dark matter' content of 20 percent, and a 'dark energy' content of 76 percent, leaving 4 percent left over for normal matter (protons and neutrons). Dark matter is invisible stuff that has

a gravitational effect, while dark energy causes the expansion of the Universe to accelerate, and has been doing so for at least the past nine billion years, since the drag of gravity lost out to it.

We have no idea of what dark matter is. About all we know is that it is traced by the normal (to us) matter of stars and galaxies, as seen in a wide-angle map derived from Hubble observations of gravitational lensing, in which intervening dark matter distorts the shapes of more distant galaxies. Nor do we have much of a clue as to the origin of dark energy. 'State of the Universe' is perhaps pretentious. 'State of Ignorance' might be better.

■ **Below:** *Newly discovered polarization of the radio waves (white lines) in this all-sky map of the Cosmic Background Radiation lead to a more precise determination of the properties of the Universe. The ripples, which represent no more than a hundred-thousandth of a degree against the 2.7 degree temperature of the CMB, are the beginnings of the clumping we see in the clusters of galaxies of today. Image courtesy NASA/WMAP Science Team.*

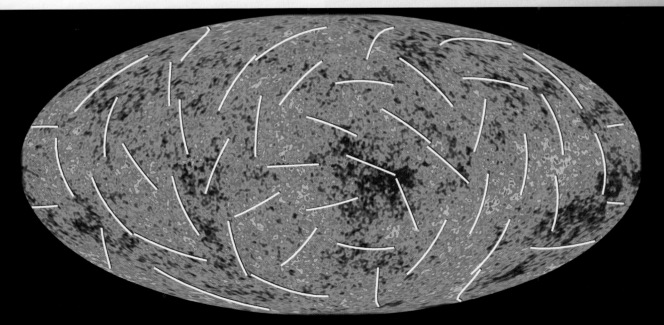

Galaxies, Galaxies, Galaxies

More galaxies are potentially visible to us than there are stars in our own Milky Way. The Sloan Digital Sky Survey has used their recession speeds to map them out to some five billion light-years, the clumps and filaments we see related in some way to the ripples fossilized in the CMB. The record actual distance, from the redshift itself, has been boosted to just shy of 7 (its light reddened by 700 percent): a 'distance' (in terms of 'look back time' into the distant past) of around 13 billion light-years. Galaxies are common out to a redshift of 6, but far less so above that, suggesting that the fountain of galaxy formation began between roughly 700 and 900 million years after the Big Bang.

Galaxy clusters are held together in part by the gravity of the mysterious dark matter. Of normal matter, the bulk is in the form of hot, X-ray-emitting gases that permeate the clusters. A mere 0.6 percent is in stars and common interstellar matter. Collisions between clusters are the most energetic events known. X-ray observations show shocks that stop the colliding hot intracluster gas dead in its tracks. Yet the dark matter revealed through lensing just keeps sailing along, demonstrating that DM is real and not an aspect of gravity that is not yet understood.

■ *Top: Hubble observations reveal that known forms of matter, stars and interstellar gases, are traced by the mysterious dark stuff. These two false-color images compare the distribution of normal matter (red, left) with dark matter (blue, right) in the universe. The brightness of clumps corresponds to the density of mass. Image courtesy NASA, ESA,and R. Massey (Caltech).*

■ *Above right: In our crowded Universe, whole clusters of galaxies can collide. X-rays reveal the collision between the hot intergalactic gases (pink), while gravitational lensing shows the dark matter (blue) that helps hold the clusters together. Images courtesy NASA/CXC/M. Markevitch et al. (X-ray); NASA/STScI; Magellan/U. Arizona/D. Clowe et al. (Optical); NASA/STScI; ESO WFI; Magellan/U. Arizona/D. Clowe et al. (Lensing Map).*

Many large galaxies, especially those that lie near the centers of massive clusters, have supermassive central black holes (bodies so compact and dense that even light cannot escape) of hundreds of millions, even billions, of solar masses. Accreting matter, including tidally disrupted stars whose debris falls into a surrounding, spinning disk, the black hole can squirt opposing jets of matter perpendicular to the disk's rotation axis across hundreds of thousands of light-years at nearly the speed of light. The hot intergalactic gas is heated and thereby prevented from falling into the member galaxies by periodic re-vitalization of the jets as the supermassive black hole chaotically consumes mass, fed through collisions and mergers with smaller systems that supply the fuel. The entire cluster and the galaxies within it are thus affected by a body not much bigger than the Solar System, testimony to their unimaginable power. At the extreme fringe are the 'quasars,' highly redshifted faint galaxies with brilliant black hole nuclei that are themselves seen to be merging as they evolve into the monsters that inhabit the massive, more mature, galaxies of today.

Our home cluster, the Local Group, is a much smaller example that contains only a few dozen members. Dominated by our Galaxy and Andromeda's M31, the Local Group is still terribly important, since we can see it in such detail. The Sloan Survey has found yet more dwarf systems, the number suggesting that there may be close to another 50. Such dwarfs may carry more dark matter per unit mass than any other kind.

The density of galaxies within clusters is so great that collisions and mergers are common. They seem to be extremely important in the basic constructions of the kinds of galaxies we see near us today. A close case is an apparent collision between Andromeda (M31) and its small elliptical companion, which seems to have punched through M31's disk, creating ripples that echo a quarter of a billion years after the event. Consistently, the central black hole of our own Galaxy (Sagittarius A') is on a par with the sparseness of our Local Group. Containing a mere two to three million solar masses, it, along with the surrounding accretion disk, is about the size of the Earth's orbit and puts out so little comparative energy that we, near the fringe of our Galaxy, are quite unaffected by it. M31's is not much different.

■ *Distant galaxies centered on each of the quasars in the first known triple system will merge to create an even larger galaxy and a bigger supermassive black hole. Image courtesy Caltech/EPFL.*

Exploding Stars

Most stars live quiet lives of non-desperation, the majority of which are spent burning (in the nuclear sense) hydrogen into helium in their deep, hot, dense cores. The higher the mass, the brighter and hotter the star. When a star's hydrogen runs out, the core contracts and heats, which makes the outer inert envelope expand in size to that of the orbits of our inner planets, and to become cooler at the surface and much more luminous. When the compressing helium core becomes hot enough, it burns to carbon and oxygen, allowing the new stellar 'giant' some rest. When the helium is gone, the core contracts again, causing the giant to grow to even larger, cooler, and to much brighter proportions. At its peak, the Sun will have expanded to the size of the Earth's orbit and will shine more than 5000 times brighter than it does today. Large size and luminosity, though, conspire to create ever-greater winds that will make our star slough off nearly all of its outer husk, leaving behind the carbon-oxygen ball. Now no bigger than Earth, it will shine forever as a cooling, dimming 'white dwarf' with a density of a ton per cubic centimeter (15 tons per cubic inch). First we fry, then we freeze.

The ageing rate depends on mass. At the low mass end of around 8 percent that of the Sun, lifetimes are far longer than the age of the Universe. (Below that limit, nuclear reactions start to shut down from lack of heat and internal pressure; such 'stars' called "brown dwarfs"). The Sun, now five billion years old, will take another five before gianthood sets in, while up in the tens of solar masses, 'hydrogen-burning' lifetimes are measured in just millions of years.

■ *Left: Hot, X-ray-emitting intergalactic gas (blue), which embeds a distant cluster of galaxies, is massively disrupted by high speed flows (observed with radio waves (red)) that pour from a billion-solar-mass black hole that lies at within the galaxy at the cluster's heart. Image courtesy NASA/CXC/U. Waterloo/B. McNamara (X-ray); NASA/ESA/STScI/U. Waterloo/B. McNnamra (Optical); NRAO/Ohio U./L. Birzan et al. (Radio).*

High mass stars, those above 8 to 10 times solar, grow not into giants, but supergiants that can rival the orbits of Jupiter and Saturn. Such stars also have such great internal energies that they can continue their nuclear progress, fusing C and O into neon and magnesium, then into silicon, then into iron. This most stable of elements can't produce energy by burning any further. Once the iron core fully develops, it catastrophically collapses from the size of Earth to that of Manhattan. Completely breaking down, the iron turns into a ball of neutrons, birthing the 'neutron star,' the density of which – 100 trillion times that of water – violently stops the collapse. The brief shocking rebound of the neutron star, perhaps aided by acoustic waves and the push of near massless neutrinos, explosively tears the rest of the star apart. In the expanding nuclear maelstrom, the 'supernova' creates all the chemical elements, including a tenth of a solar mass of new iron. Core-collapse supernovae are one of the three major sources of the chemical elements other than the hydrogen and helium that were born in the Big Bang.

The other two sources are nuclear-enriched winds from giants and a second kind of supernova. White dwarfs are supported by electrons that under high-temperature conditions have been stripped from their parent atoms. Since electrons behave like waves as much as they do particles, they cannot be pushed any closer. But get the white dwarf above a critical limit of 1.4 solar masses, and even the electrons can't do the job. If enough fresh matter from a tidally disturbed binary companion falls onto the white dwarf to push it over the limit, it collapses and explodes even more violently than the core-collapse version. These 'Type Ia' supernovae (the other kind thus 'Type II') are so uniform and brilliant that they are crucial in the study of the acceleration of the Universe's expansion. If it all sounds understandable, it isn't, since the exact explosion mechanisms of both kinds yet elude us.

Historical supernovae (six in our Galaxy in the past millennium) plus radioactive debris in the form of interstellar radioactive aluminum suggest that supernovae occur at a rate of one to three per century (a good fraction hidden behind dust clouds). Given the rate and the age of Earth, there's a good chance that every quarter-billion years or so, one will go off within a 'death zone' of 30 light-years. Such an explosion would not only illuminate the night with the light of 100 to 300 full Moons, but would strip the ozone layer, allowing harsh daytime solar ultraviolet light to come pounding down on us, destroying the food chain and leading – perhaps – to a mass extinction. The discovery of radioactive iron in sea beds tells us that it has indeed happened. More, discovery of the by-products of radioactive supernova debris in meteorites indicates that one went off close to the Sun even as our Solar System was being formed.

■ Core-collapse Supernova 1987A blew up 20 years ago in the Large Magellanic Cloud (a nearby satellite galaxy). Its brilliant energetic light first illuminated the hourglass-shaped envelope lost by the progenitor supergiant. The shock wave then proceeded to pound the inner edges of the envelope to create a brilliant "string of pearls" that is a light-year across. The neutron star or black hole that should be at the center has never been found. Image courtesy NASA, ESA, P. Challis and R. Kirshner (Harvard-Smithsonian Center for Astrophysics).

Little Stars

White dwarfs may be small, but they are vast compared with neutron stars. What they lack in size is compensated by other amazing properties, not the least of which is density. As iron cores collapse, they spin faster until they are rotating up to dozens of times per second. Contraction of the original star's magnetic field coupled with that generated by spin yields lethal fields a trillion times that of Earth. The rapid rotation generates radiation that spills out along a tilted magnetic axis, which wobbles wildly around as the star spins. If the Earth is in the way, we see a 'pulse' of light or radio waves, hence the set of neutron stars called 'pulsars.' We've so far discovered around 1800 of them.

■ NGC 2440, a planetary nebula a light-year or so across, expands around a new, very hot white dwarf (the tiny dot seen at the center), the just-revealed carbon-oxygen core of what was once a giant star. The nebula fluoresces under the harsh ultraviolet light from a stellar surface heated to 220,000 Kelvin, nearly 40 times hotter than the Sun. Image courtesy NASA, ESA, and K. Noll (STScI), and the Hubble Heritage Team (STScI/AURA).

Pulsars start life radiating across the spectrum. As they radiate, they slow down, continuously dropping their highest-energy radiation until nothing is left but radio, then finally go quiet. There must be 100 million or more peaceful neutron stars roaming the Galaxy. Yet we find that some, perhaps most, do not conform to the classic picture. A few produce short radio bursts that are among the brightest things ever seen, while a perverse slow pulsar that should not exist takes a leisurely 6.7 hours between pulses. Weird ones keep sliding home, showing how little we really know about them.

Gamma-Rays

Several times a day, orbiting satellites record intense bursts of gamma-rays (GRBs) coming at us from all over the sky. There are two kinds, long and short, separated at about two seconds duration. The long ones, promptly followed by powerful optically-visible afterglows, have been related to brilliant core-collapse supernovae – *hypernovae* – in ultradistant galaxies. Their power and rarity per galaxy suggest that they derive from equally-rare rapidly rotating massive stars that have lost much of their outer layers. Spinning ever more quickly, the cores collapse into black holes. As they go down into the pit, the outgoing explosions become focussed perpendicular to the rotation into twin opposing jets whose shock waves create the gamma-ray bursts. The emerging energy hitting the surroundings then makes the afterglows. If we are in the line of fire, we see the GRB; otherwise nothing but the hypernova.

The hypernova's extreme power expands the death zone around Earth into the hundreds, even thousands, of light-years. It's suggested that at least one of the 'big five' mass extinctions the Earth has suffered over the past billion years resulted from one. We are saved, however, first by the beaming (which would have to be directed at us) and also by GRBs seeming to prefer small galaxies with low metal contents, not those like ours. Our best hypernova candidate is the southern hemisphere's Eta Carinae, which consists of two advanced-state 'hypergiants' born with over 100 solar masses each, the leader of which could go off anytime. Fortunately, it's 7500 light-years away, and its deadly rotation axis points elsewhere.

Then there are the less-energetic short GRBs to worry about. Since stars like to double up, it makes sense that their end-products do too. Intimate binary neutron stars, which severely distort their local spacetime, would radiate gravity waves that would slowly draw them together. Violent mergers between the pair (or between neutron stars and black holes) seem to be the best candidate for the short bursts.

Even single stars can produce gamma-ray bursts. At the high end of the neutron star ranks are 'magnetars' that may also be products of hypernovae and that have magnetic fields a hundred to a thousand times greater than those of normal pulsars. Neutron stars are encased in solidified crusts that must occasionally re-adjust

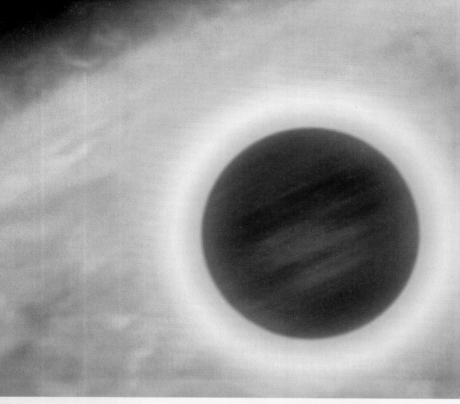

themselves as the rotations slow down. Pulsars respond by briefly increasing their rotation speeds. Magnetars of the rare 'soft gamma-ray repeater' variety do it by blasting huge bursts of gamma-rays. The 1998 burst from SGR 1900+14 disrupted Earth-orbiting satellites and ionized the upper atmosphere from a distance of 20,000 light-years, and the 2004 blast from SGR 1806-20 topped that by a factor of 100 from more than twice that distance. Identified in nearby galaxies, magnetars thus make at least a part of the population of the short bursters.

Just Plain Stars

As ageing giants lose their outer envelopes through winds, the exposed cores that are becoming white dwarfs light up the fleeing gas to produce beautiful (but mis-named) 'planetary nebulae.' Binary-star action seems to create some of their bizarre shapes, but so might magnetic fields. Mass loss is epidemic. Even Cepheid variables, our primary 'standard candles' for getting distances to other galaxies, seem to be surrounded by dusty cocoons.

In other 'home town news':

■ *RS Ophiuchi, a repeating nova (caused by matter being dumped by a companion onto an orbiting white dwarf whose surface then explodes), went off again;*

■ *Massive open clusters are still forming in the Galaxy;*

■ *Vega is a fast spinner whose axis points at us and like all such stars is hotter at the equator than at the flattened poles, which solves long-standing problems;*

■ *Polaris (a subtle Cepheid) has another companion, making it a triple star;*

■ *Proxima Centauri really <u>does</u> belong to Alpha Centauri;*

■ *There may be as many as 100 billion brown dwarfs in the Galaxy, raising the number of its stars to 300 billion.*

Birth

Stars are born from condensing blobs within the cold dusty molecular clouds of interstellar space. Initial compression is supplied by shock waves induced by massive star winds and supernovae, star death creating not just the chemical elements of life but the new stars that host life itself. New stars form dusty circumstellar disks from which they accrete additional matter and that ultimately condense into planets. At least they do unless the disks are destroyed by intense radiation from nearby massive stars, surely one of the great ironies of nature, since massive stars are *required* for star formation through their compression-triggering supernovae.

Barring disruption, planets are thus natural by-products of star birth. Over two hundred have been found with the standard radial velocity technique (the planets shifting their parent stars back and forth), including 25 multiple-planet systems. Mu Arae tops the list with four. Several more have been found by planets transiting in front of their stars. We initially expected other planetary systems to be like ours. Instead, there is a rich variety that includes Jupiter-like bodies tucked up next to their stars in extremely short-period orbits, planets with highly eccentric orbits, planets in double star systems, a giant star with a planet (Pollux), and 'puffy jupiters' that are too big for their masses. The general rule is that stars with planets are richer in metals than the Sun. Yet cool, low mass stars with planets confusingly seem to have lower metal contents. We are a long way from sorting it all out, no surprise since we have not been observing extrasolar planets long enough to detect the full range of possibilities.

Star and planet-forming clouds have a rich chemistry that may find its way into the birthing planets and that may even form the seeds of life. With the discovery of radio radiation from eight new interstellar molecules that include complex organics, the number known rises to a total of 141. Only 50 years ago, we had just two.

There is even room for new atomic discoveries. The Galaxy has long been mapped by the radio emission of ordinary hydrogen at a wavelength of 21 centimeters. The analogous emission from deuterium, hydrogen's heavier isotope (a neutron added to the proton) at 92 centimeters has finally been detected. Natural deuterium was created in the Big Bang. The amount is right on the prediction derived from the ripples in the Cosmic Microwave Background, and we are back to the top of the story, which like star birth and death is cyclical, no part in ascendance, each part of vital importance to all the others.

■ *Stars, even whole clusters of stars, are born from the dusty gases of interstellar space. High mass stars, which blow bubbles in the surrounding interstellar matter, and their resulting supernovae act to create yet new stellar generations. Image courtesy NASA, ESA, and the Hubble Heritage Team (STScI/AURA) - ESA/Hubble Collaboration.*

*James B. Kaler, **Professor Emeritus of Astronomy at the University of Illinois, a student of stars and planetary nebulae, has written several books on astronomy. His most recent is The Cambridge Encyclopedia of Stars, published by Cambridge University Press.***

4
High-Stakes Astronomy AT LOW FREQUENCIES...

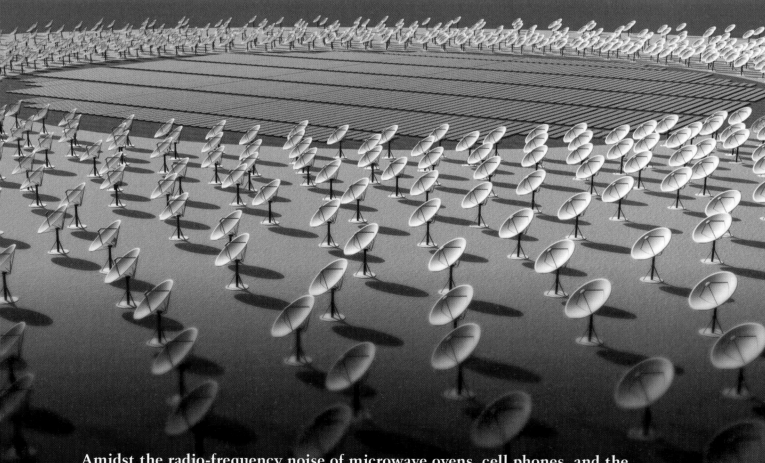

Amidst the radio-frequency noise of microwave ovens, cell phones, and the occasional unfiltered stereo system, are astronomical features of our universe waiting to be found. In this article, *Carolyn Collins Petersen* conveys the enormous lengths to which astronomers are going in order to extract astronomical

...Opening Up a New

WAVELENGTH FRONTIER

THE UNIVERSE is an equal-opportunity radio-frequency emitter. Point a dish (or a radio detector array) in any direction and you can find something interesting to study. This is because anything active in the universe gives off a range of wavelengths and frequencies, including low-frequency signals below 400 megahertz (MHz). This includes our Sun, which puts out radio signals across a broad range of frequencies. The planet Jupiter does interesting things that we can study between 15-30 MHz. Exploding stars and the center of the Milky Way Galaxy are also among the millions of sources that astronomers have observed at low frequencies. Yet, until very recently, this range of the radio spectrum remained largely unexplored by astronomers because signals were difficult to detect through all the noise of our technology.

Did you talk on a cell phone today? Use a wireless computer? Transfer some money from checking to savings using an ATM machine? Listen to a radio broadcast in your car? Fly in a plane? If so, you used technology that depended at some point on the use of radio waves. Our lives are filled with gadgets that use radio waves, but these pieces of modern technology also put out radio frequency interference (RFI). This 'noise' can and does interrupt or completely mask signals from space. Sifting them out from interference by both natural and human-made sources is a challenge.

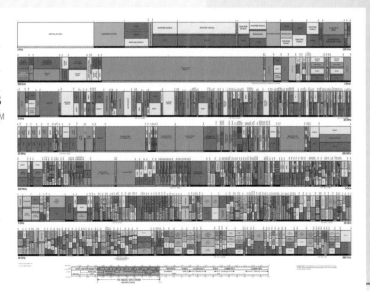

■ *Left: Frequency allocations in the radio spectrum cover everything from 10 kilohertz (very low frequencies, LHF) to 30 gigahertz and above (extremely high frequency, EHF) wavelengths. All radio-emitting technologies, from cable-locating equipment and power line carrier systems to aircraft navigation systems and industrial/business uses are shown on this spectrum chart. Radio astronomy peeks through at various wavelengths. Unfortunately, some of the most interesting and unexplored parts of the radio astronomy spectrum lie in the lower end of the spectrum, at frequencies up to a few hundred megahertz. These same frequencies are heavily populated by telecommunications and broadcast transmissions, which can partly or completely block off-planet signals. Image courtesy National Telecommunications and Information Administration, U.S. Department of Commerce.*

Frequencies and Wavelengths

In radio astronomy, signals are often stated in terms of frequency and/or wavelengths. Frequency is the term for the number of cycles a radio wave goes through per unit of time. You often see the terms 'megahertz' or 'kilohertz' used in relation to radio waves. One hertz (often abbreviated Hz) is one cycle per second. Since radio waves go through hundreds, thousands, millions, or billions of cycles per second, metric prefixes are used to describe those frequencies. Kilohertz is abbreviated kHz and means 'thousands of hertz'; megahertz (MHz) is millions of cycles, and gigahertz (GHz) is billions of cycles.

For purposes of low-frequency astronomy, astronomers concentrate on a frequency range between 10 and 400 MHz. For comparison, the radio wavelengths covered by all of radio astronomy range from 10 to 10,000 MHz, with VLBI experiments extending the range up to 245 GHz and the range of the Smithsonian Submillimeter Array topping out at around 800 GHz. The Atacama Large Millimeter/submillimeter Array, which will be exploring the 'cool' universe (radiation leftover from the Big Bang, molecular gas and dust clouds, planetary systems, and galaxies) will be sensitive at submillimeter wavelengths and frequencies between 80 and 900 GHz.

In addition, you can express a radio signal in terms of its wavelength (literally, the length of the wave). A longer wave is a less-energetic signal; a shorter wave carries more energy. To derive a signal's wavelength in meters from a frequency, use the following formula: 300/frequency (in MHz).

■ **Above right:** *Low-band (top) and High-band (middle) antenna elements in the Low-Frequency Array (LOFAR), located in the Netherlands. The Low-band antenna is sensitive to frequencies between 30-80 MHz. The High-band antenna is sensitive to frequencies between 120-240 MHz. These detectors should be able to trace signals from such celestial events as supernova explosions, black hole collisions, and signals from the earliest epochs after the Big Bang. The data will be processed in the STELLA supercomputer (bottom). Images courtesy ASTRON, LOFAR Consortium.*

The Low-Frequency Astronomy Race

What kinds of signals are we talking about? Currently, radio astronomers are racing to the low-frequency end of the electromagnetic spectrum to search out everything from the oscillations of primordial hydrogen atoms that existed when the first stars began to shine, to the blasts of radio noise emitted when stars and galaxies collide. It also turns out that some distance-stretched signals from other civilizations, if they exist, could be hidden 'in the noise' of the low-frequency spectrum.

All of these signals are in plain 'view' of a new breed of existing and planned radio arrays specifically sensitive to low-frequency radio signals that lie in sections of the frequency spectrum that we also use for radio, TV, military and civilian communications. The installations, such as the Very Large Array in Socorro, New Mexico, contain dozens to thousands of detector antenna units. They employ interferometry to gather a good 'picture' of an object in the sky.

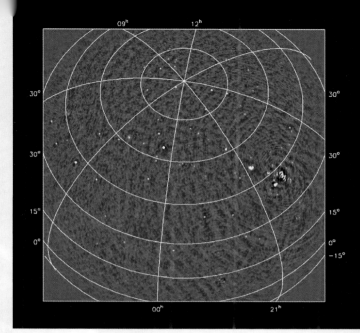

Interferometry combines and correlates data from the telescopes in an array to form multiple pixels of the sky. The pixel size determined by the size of the baseline between array elements.

Still, just to have a ghost of a chance of reliably detecting astronomically meaningful low-frequency signals buried by our radio and TV broadcasts, at least two of these observatories – the the Mileura Wide-Field Array (MWA) and the Square Kilometre Array (SKA) – are being planned for installation in very radio-quiet areas in the southern hemisphere. Australia is hosting MWA, and as of this writing, the SKA consortium is considering Australia and South Africa for the final site.

While you may think that all radio arrays are 'out there' in the desert somewhere, like the MWA, or the Very Large Array, several of these detector collections are sited near local farms, or at observatories nestled up to people's backyards. Operators of these systems, such as the Low-Frequency Array (LOFAR) in the Netherlands, maintain they can also do the same science as their more distant counterparts, but to achieve it, they must develop sophisticated algorithms and filtering techniques to tease out the low-frequency whispers of exciting cosmic events.

An Array in My Corn Field?

What happens when an array must be built in a populated place (for political, financial, or scientific reasons)? It doesn't automatically mean that science can't be done. This is a challenge the builders of LOFAR (currently under construction in the Netherlands), are working to overcome in their detector designs. This array is under located on farmlands and open spaces across the Netherlands, and will ultimately spread 15,000 antennas distributed over 77 stations across that country and possibly into Germany. When fully deployed, LOFAR could have a maximum baseline between its most distant elements of about 360 kilometers, and will be sensitive to frequencies between 30 and 240 MHz. The interior portion of the array will be able to look in as many as 200 directions simultaneously. The antennas will pump data to collector sites, then to a series of remote stations, and ultimately to a core computing center where systems will correlate the data and send it along to scientists for study.

The builders of LOFAR and other arrays close to human habitation must operate in what its designers call a 'hostile RFI environment.' For LOFAR, mitigating interference by using filtering technology will be a key to success. The first array elements are in place, and in 2006, LOFAR scientists and engineers began reporting test results.

Deuterium from the Backyard

Advances in filtering technology for low-frequency radio astronomy got a demonstrable boost in 2005 when the operators of a series of antennas called the Deuterium Array, located at Massachusetts Institute of Technology's Haystack Observatory in northern Massachusetts, reported the detection of oscillations of deuterium atoms at 327 MHz. This signal is extremely sensitive to interference, and so in order to 'see' deuterium at that wavelength, the receivers had to be shielded from outside RFI.

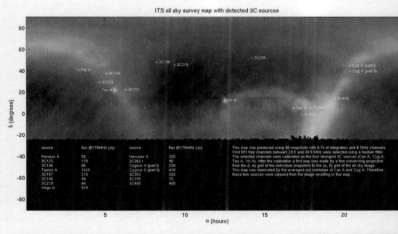

■ **Top:** *The first high quality deep wide-field image with a LOFAR station, released 25 April 2007. The data were collected with 96 low-band antennas located in four fields at the heart of the array in the province of Drenthe in the North-East of the Netherlands, and transported over a dedicated glass-fiber link to a central processing facility at the University of Groningen, some 60 km away. The image was made at a frequency of about 50 MHz and is centered on the bright radio source Cassiopeia A. At least 40 other sources can be seen in this image. Image courtesy ASTRON, LOFAR Consortium.*

■ **Above:** *Early results from the LOFAR ITS test site show a map of the sky as seen between 29.5 and 30.5 MHz. Some easily recognizable structures, such as the Perseus A region, quasar 3C123, the Cygnus X object, and other quasars and the cores of star-forming regions stand out at these wavelengths. Image courtesy Stefan Wijnholds (ASTRON).*

As Massachusetts Institute of Technology scientist Alan Rogers and his team of researchers built the Deuterium Array, they rose to the challenge of weeding out radio noise by sniffing out sources of RFI from nearby houses in the neighboring towns of Westford and Groton. By replacing such things as answering machines and stereo parts to make a quiet environment, they cleared the way for the array to search out the signals from primordial deuterium located at the 'anti-center' of the Milky Way Galaxy. The practice paid off. After a year of data-gathering, the team reported the first unambiguous detection of deuterium at 327 MHz, in a paper by Rogers et al. entitled 'Deuterium Abundance in the Interstellar Gas of the Galactic Anticenter from the 327 MHz Line' published in the 1 September 2005 issue of *Astrophysical Journal*, one of the 'bibles' of astronomy science results.

What is deuterium and why all the fuss over detecting it at the low end of the frequency spectrum? Deuterium is an isotope of hydrogen, and most of the naturally occurring kind was created in the Big Bang, some 13 billion years ago. It's something of a 'fragile' element because deuterium can be destroyed very easily in stellar interiors, for example, or by the heat generated in the starbirth process. Areas in galaxies where star formation is active (or has been in the past) do not have very high amounts of primordial deuterium. Thus, deuterium (or the lack of it) is what scientists call a 'tracer' of stellar activity.

The Deuterium Array specifically zeroed in on a region of the Milky Way Galaxy where the scientists expected that some primordial deuterium would still exist in the interstellar gas. They focused on the 327 MHz transition of deuterium because it allows them to measure the relative amounts of deuterium to hydrogen (called the D/H ratio) and apply it to sources anywhere in our galaxy. A small amount of deuterium relative to hydrogen means that most of the deuterium that may have existed in the area has been destroyed or locked in a chemical bond with another element. The Deuterium Array result showed a D/H ratio of 23 parts per million in the region of the galactic anti-center (a region well away from the center of the galaxy, where relatively little star formation has taken place).

■ **Above right:** *Two views of the MIT Haystack Observatory deuterium array. This was an electronically steerable multi-beam array of twenty-five 5 x 5 crossed dipole stations sensitive to deuterium emissions at 327 MHz. It performed its tasks within a few hundred feet of a multitude of houses in Westford and Groton, Massachusetts, communities about 30 miles northwest of Boston. The small silver trailer (arrowed in the aerial photograph, and also shown in the close-up view) is a radio frequency interference monitor that kept track of stray radio noise during the array's operation. Images courtesy MIT Haystack Observatory.*

In larger cosmic terms, astronomers have known for some time that the amount of deuterium in the universe has a bearing on how much regular matter (called 'baryonic' matter) there is, as well as how much photonic 'matter' (light) there is. Because of the way it was created in the Big Bang, accurate measurements of deuterium allow astronomers to understand more about the mechanics of creation. In addition, knowing how much deuterium existed both then and now will let astronomers calculate the density of cosmic baryons (regular matter) in the universe. That density would indicate whether ordinary matter is bound up in black holes, gas clouds, or brown dwarfs, or is more luminous and tied up in making stars.

However, there is a substantial amount of something else in the cosmos that we can't detect (yet), but we can sense its gravitational pull on ordinary matter. It's called dark matter, and measurements of deuterium at as many wavelengths as possible will also give scientists a handle on just how much dark matter there is 'out there.' Thanks to the Deuterium Array, they are a few steps closer to that goal.

An Array Out Back in the Outback

It may seem an unlikely place for radio astronomy, but a livestock station in far Western Australia is the site of another unique array that is also under construction, but is already studying the universe at frequencies below 300 MHz. Unlike the LOFAR array sited in a radio-noisy part of Northern Europe, the Mileura Wide-field Array takes advantage of one of the few remaining radio-

4 - High-Stakes Astronomy at Low Frequencies . . . Opening Up a New Wavelength Frontier

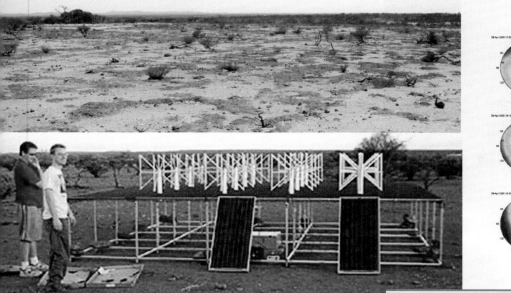

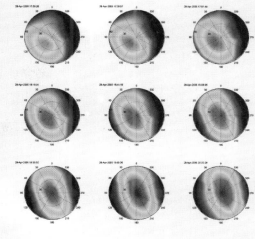

quiet areas on the planet, a region of Australia nearly 200 miles (300 kilometers) inland from the country's western coast and safely located well away from major cities and sources of radio frequency interference. It has excellent sky access, especially to the center of our Milky Way, and the Large and Small Magellanic Clouds (our galactic neighbors). The array is being built in two parts: a Low-Frequency Demonstrator (LFD, active in the 80-300 MHz range) and the New

■ **Top left:** Outback Australia, at the Mileura Wide Field Array site, is the sort of rugged, radio-quiet region where low-frequency astronomy can be carried out more easily. Image courtesy Colin Lonsdale, MIT Haystack Observatory.

■ **Bottom left:** Students help construct a prototype of the 500 tiles of the MWA in Australia. Each tile has 16 antennas sensitive to frequencies between 80 and 300 MHz. Image courtesy Merv Lynch, Curtin University of Technology, Australia.

■ **Above:** An image sequence showing the galactic center passing overhead at Mileura. The frequency is 108 MHz and the images were made by steering a single-tile beam to 40 grid locations in rapid succession. The zenith is at the center of the image. Image courtesy Judd Bowman, MIT Haystack Observatory.

Low-Frequency Astronomy Targets

The ambitious science programs that low-frequency astronomers want to pursue echo a broad range of astronomical topics. Examine the web sites for the science goals of the MWA, SKA, and the Allen Telescope Array (planned for full deployment in California), and the United States Navy's Low-Frequency Array, and you'll see many of the same science targets. Let's examine some of them in more detail.

First, there's the Epoch of Reionization (often called EOR, for short). This is something of a Holy Grail at nearly all frequencies. It's a period early in the universe about a billion years after the Big Bang when the first luminous sources (such as stars, galaxies, or quasars) began to light up the intergalactic medium (filled with neutral hydrogen gas). By studying the EOR at many frequencies, astronomers can learn about how structures like galaxies and galaxy clusters formed in the tumultuous times after the Big Bang. It casts insight into the distribution of matter and how today's highly structured universe may have evolved. Low-frequency detection of events at these early epochs depends on being able to sift out the signal of what astronomers call 'the 21-cm hyperfine transition line of neutral hydrogen' redshifted to frequencies below 200 MHz. Theoretically, low-frequency arrays that can detect this line will be able to probe the reionization of the early universe in great detail, giving astronomers a feeling for the density, temperatures, and velocities of material that existed back then. A wide field of view, as possible from an array, would show huge parts of the early universe.

Second on everybody's hit list of cosmic study targets are transient radio sources, the so-called 'transient sky.' These are things like cosmic rays, pulsars, gamma-ray burst afterglows, the aftermath of supernova explosions, the huge bursts of energy released when massive objects (like black holes or neutron stars) collide, and the radio emissions from extrasolar Jupiter-like planets.

Next are low-frequency surveys of distant galaxies to count and classify them, followed by studies of the molecules in the interstellar and intergalactic mediums. The abundances of certain molecules, such as oxygen, hydrogen, carbon, and compounds made with combinations of those elements, reveal a great deal about the processes that created them, such as starbirth and stardeath.

Planetary science is also a big part of any program, including the study of possible naturally occurring low-frequency signals from Jupiter-sized (and larger) exoplanets with strong magnetic fields. In our own neighborhood, solar activity spurs a tremendous amount of space weather - defined as the events that occur as a result of interactions between material ejected from the Sun and our planet's upper atmosphere and extended magnetic field.

An artist's impression of an Earthlike planet with a radio-loud civilization. If communicative life existed on another planet, it's possible that some portion of its radio communications could be detected by low-frequency-sensitive arrays on Earth. Image courtesy David A. Aguilar, Harvard University Center for Astrophysics.

of CFA) suggest that such accidental leakage from military radars, as well as broadcast TV and radio on another planet would be detectable with the MWA-LFD. A SETI program using the array could detect Earth-like radio signals from civilizations up to 30 light-years away during a one-month-long staring session. If these signals were found, then the arrays (along with other SETI surveys) could make more observations to measure the rotation rate of the source planet and the length of its year.

The MWA is being created in a partnership between institutions in Australia and the United States, including the Massachusetts Institute of Technology, the Harvard-Smithsonian Center for Astrophysics, the Australia Telescope National Facility, the Australian National University and Curtin University, and the Government of Western Australia. Funding also comes from the U.S. National Science Foundation. The Low-Frequency Demonstrator and the New Technology Demonstrator will start extensive operations in 2008. A second stage of the project, with more sensitivity and resolution will follow the successful deployment of LTD and NTD.

Other institutions from Europe and Asia are also cooperating in using and developing the LFD, which has been tested in the field and is now taking data. The current collaboration is focused on measuring interplanetary scintillations, radio bursts, and ionospheric perturbations caused by coronal mass ejections from the Sun. These are aimed directly at understanding space weather phenomena and how they can affect life on our planet.

Technology Demonstrator (NTD, sensitive to frequencies between 800 and 1600 MHz).

The MWA-LFD is designed to detect and characterize highly redshifted 21-centimeter emission from hydrogen molecules in the early universe. One of its most important scientific goals is to create a three-dimensional map of ionized 'bubbles' that formed as the first quasars and galaxies flooded space with ultraviolet light billions of years ago. Along with the NTD, which is also serving as a test bed for technology to be used in the upcoming Square Kilometre Array (SKA), MWA will track transient radio sources such as pulsars, low-frequency signals from molecules in the interstellar medium, and propagation effects due to the Sun's influence on our planet's upper atmosphere.

Eavesdropping on the Universe

An interesting use of the Low-Frequency Demonstrator was proposed at the 2007 American Astronomical Society meeting in Seattle, by theorist Avi Loeb of the Harvard-Smithsonian Center for Astrophysics. He argues that while most Search for Extraterrestrial Intelligence (SETI) programs are looking for specific signals deliberately beamed across space, searchers should also be on the lookout for signals that leak from regular communications channels. Loeb and co-author Martias Zaldarriaga (also

SKA and the Future of Large Arrays

The future of large arrays is inarguably the Square Kilometre Array (SKA), to be located in either Australia or South Africa. Its frequency range is 100 MHz to 25 GHz, so it will be covering part of the low-frequency radio spectrum. SKA is designed to complement the Atacama Large Millimeter Array (located on the Atacama Desert of Chile), and the James Webb Space Telescope, the orbital follow-up to Hubble Space Telescope, and planned to be mostly sensitive to infrared light with some visible-light capability. Both of these are currently in development for use within the next decade.

SKA will consist of several thousand antennas combined to give a radio 'aperture' equivalent to a million square meters of collecting area with a sensitivity nearly a hundred times that of existing large arrays! The exact configuration is still under design, but one possibility is for 30

stations, each with a collecting area of a 200-meter telescope, and another 150 stations each with the collecting area of a 90-meter telescope. This would give the array an incredibly high-resolution view of the low-frequency universe.

Development of SKA began in 1991. It is still in the planning stages, including the technology test bed in Australia. The international SKA consortium (which includes more than 50 institutions in 17 countries) oversees all science and technical developments. It makes all decisions on location and funding, as well as construction. If all goes well, construction could start as early as 2011.

What will SKA look at? As with the other arrays, this huge installation will focus its highly tuned radio sensors on the big picture – literally! Its planners expect to make new discoveries in astroparticle physics, cosmology, fundamental physics, galactic and extragalactic astronomy, and solar system science. Like its forerunners, SKA will focus on the Epoch of Reionization, using what it finds to understand how the first galaxies assembled themselves, how early black holes formed, and how these objects influenced the primordial cosmic environment. Other topics for study include the origin and evolution of cosmic magnetism, the search for astrobiological precursors to life and earth-like planets, and the evolution of galaxies and the large-scale structure of the universe.

With their high resolution, advanced filtering technologies and/or radio-quiet locations, coupled with extremely high data rates and computerized correlation, the low-frequency observatories in existence now and those planned for the near future are giving astronomers unprecedented chances to study hundreds of square degrees of sky at difficult-to-observe frequencies. These arrays are peering across time, into places we could never see, at astrophysically important signals from objects that give off some the most fleeting and hard-to-observe flickers of radio waves in the universe.

For further information, consult these websites:

Allocation of radio spectrum:
http://www.jneuhaus.com/fccindex/spectrum.html

Background on radio astronomy in general:
http://www.haystack.edu/edu/pcr/resources/BasicResources.html

ALMA web site:
http://www.alma.nrao.edu/

Haystack Observatory Radio Arrays web site:
http://www.haystack.edu/ast/arrays/index.html

LOFAR web site:
http://www.lofar.org/

SKA web site:
http://www.skatelescope.org

Carolyn Collins Petersen is a Massachusetts-based science writer specializing extensively in astronomy and space science. She is first author on **Visions of the Cosmos** (co-authored with John C. Brandt), a book that explores the multi-wavelength universe. She was the senior science writer for the Los Angeles, California-based Griffith Observatory astronomy exhibition (which opened in late 2006), and is currently working on a series of vodcasts about space weather, and documentary scripts for planetarium and science center use.

5

Where Do All the Stars COME FROM?...

During the past few years our knowledge of how stars form, and their intimate relationship with dusty disks that play an integral part of star and planet formation, has leapt ahead at a dramatic pace. Join star formation experts, *Luisa Rebull* from the Spitzer Science Center and *Stephen Strom* from the National Optical Astronomy Observatory, in a unique and intriguing insiders' view of our current state of knowledge from the researchers themselves.

Image courtesy NASA/JPL-Caltech/N. Flagey (IAS/SSC) and A. Noriega-Crespo (SSC/Caltech).

...New Views of Star Formation
WITH THE
SPITZER
SPACE TELESCOPE

VAST CLOUDS of gas and dust are swirling throughout our Milky Way galaxy. Many of these clouds are stellar nurseries, places where one star (in the case of small clouds) to tens of thousands of stars (in the case of the largest and most massive clouds) are being born right now. These clouds range in size from cores that are 100,000 times the size of the Solar System and a mass of several suns (solar masses), to giant clouds more than ten million times the size of our Solar System and many thousands to tens of thousands of solar masses. A typical star-forming cloud might create very few massive stars (20 solar masses or more), many stars like our Sun, and many more lower-mass stars and brown dwarfs, which are objects with a mass smaller than the 0.08 solar masses needed to produce stars fueled by nuclear fusion. An umbrella term for all of these newly-forming objects is young stellar objects (YSOs).

Astronomers are beginning to use the power of the Spitzer Space Telescope, combined with other telescopes, to understand many things about star formation in our Galaxy. (See article 'Detecting Heat Radiation from Space ... The Spitzer Space Telescope' by Michelle Thaller in last year's volume, *State of the Universe 2007*.) These include the factors that control the efficiency of the star-formation process in different clouds, what determines whether a high or low mass star forms in different parts of stellar nurseries, and how the process of star formation across all masses is related to the formation of planetary systems and ultimately life-bearing planets analogous to our own Solar System.

Astronomers have developed guidelines – reasonable ideas – regarding the basic processes of star formation for stars like our Sun. This working framework derives both from numerical modeling and observations of forming YSOs – not only with Spitzer, but with other telescopes that are capable of probing the physical, chemical and dynamical state of the gas and dust that ultimately collapses to form stars. Similar processes are likely to operate in much more massive stars (only faster), and brown dwarfs (only slower).

Visible   *Infrared*

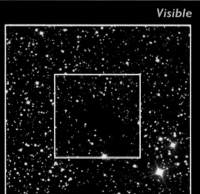

"Starless" CoreL1014

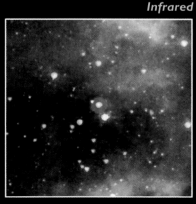

Spitzer Space Telescope IRAC MIPS
Visible DSS

■ **Left:** *Comparison of visible (left) and Spitzer infrared (right) image of the protostellar core L1014. The bright yellow object at the center of the right-hand image is a forming star, detected via Spitzer's ability to penetrate the optically opaque dust contained in protostellar core. The red ring surrounding the object is an artifact of the reduced spatial resolution of the telescope at 24 microns. Images courtesy NASA/JPL-Caltech/N. Evans (Univ. of Texas at Austin)/DSS.*

■ *HH 46/47, a YSO ejecting a jet and creating a bipolar outflow. The central protostar lies inside a 'Bok globule' hidden from view in the visible-light image (inset, lower left). Image courtesy NASA/JPL-Caltech/A. Noriega-Crespo (SSC/Caltech), Digital Sky Survey.*

Stars are born in rotating cores, comprised of roughly 100 parts gas and 1 part dust. These begin to collapse when the force of gravity (which is dependent on the mass and radius of the core) exceeds the internal pressure (a measure of the internal motions of material within the core – motions that push back against the inward pull of gravity). The collapsing, rotating core forms a central 'stellar seed' surrounded by a geometrically thin disk. The seed-disk system is continually fed from the material in the rotating core. The orbiting disk material plays a crucial role in two ways: first, in transporting material from the core to the central seed, eventually (over several hundred thousand years) allowing the seed to reach a mass comparable to that of the Sun; and second, in providing the material from which planets can form.

These optically-obscured objects emit the majority of their light in the infrared. Prior to Spitzer, we knew of only about 50 stars at this stage of development. However, with Spitzer, the number is dramatically higher. We have now detected several hundred of these protostellar cores, whose presence is inferred from unique patterns of brightness versus wavelength. At Spitzer's shortest wavelength of 3.6 microns, the light comes mainly from the newly-forming star at the heart of the core. At longer wavelengths, the light from the object

becomes stronger, one signature that it is not a background star. Also, at the longer wavelengths of 8 and 24 microns, you can see the glow from interstellar dust surrounding the YSO (glowing green to red in the Spitzer composite image of the L1014 protostellar core). This dust consists mainly of a variety of carbon-based organic molecules known collectively as polycyclic aromatic hydrocarbons (PAHs). The red color traces a cooler dust component.

Owing to its tremendous sensitivity, Spitzer is able to probe star-forming clouds more than 20 times further from Earth than had been probed by previous ground- and space-based observations. As a result, we now have a catalog of newly-formed stars spanning nearly the full range of known stellar masses – a crucial precursor to understanding what kinds of natal protostellar cores give birth to what kinds of stars and why. Follow-up observations with sensitive radio telescopes provide measurements that allow us to assess the masses, internal motions and rotation speeds of these cores – critical factors to understanding the kinds of conditions conducive to forming stars of different masses, and to determining the characteristics of the disks that ultimately form planetary systems.

The birth of a star is a violent event. Early in its evolutionary history, while the YSO is still embedded and optically invisible, a powerful, highly-collimated, jet-like outflow begins to emerge from the protostellar core. Outflows are a signpost that material from the core has formed a disk surrounding the embedded stellar seed, and that the disk has begun to transport material to the surface of the seed. This accretion process results in the launching of jets, or outflows. The precise launching mechanism for jets is unknown, but is likely to be related to rotation, magnetic fields, accretion, and the interaction between the star and the disk. About 10 percent of the matter that is accreted through the disk is ejected as an outflow or jet. The jets can reach sizes of trillions of kilometers and velocities of hundreds of thousands of kilometers per hour.

The launching of jets through the accretion process may be nature's way of limiting or controlling the ultimate mass of the star that forms as a result of their disrupting the core and removing material that would otherwise ultimately reach the surface of the core. With Spitzer, the star and its jets of molecular gas appear with clarity. The 8-micron channel is sensitive to emission from PAHs, which are excited by the surrounding radiation field and become luminescent. The shock fronts from jets of material plowing supersonically into the cold, dense gas nearby are easily visible. In particularly active regions of star formation, such as that seen in NGC 1333, hundreds if not thousands of jets are found, each pair of which can be traced back to a parent YSO.

During the assembly of the star-disk system, both the central star and disk are obscured from observations using conventional optical telescopes by the collective effects of micron-size dust grains in the natal material surrounding the YSO. Through the combined effects of incorporation of core material onto the star and the dissipative role of outflows/jets, the cocoon of material surrounding a forming YSO is disrupted, ultimately rendering star-disk systems visible. Superb examples of this phase of star formation are provided by Hubble Space Telescope imaging of nearby star-forming regions such as the Orion Nebula Cluster. Here, the background light from the Orion Nebula (an ionized hydrogen region) allows the emerging star-disk systems to be seen in silhouette, or glowing themselves in response to ultraviolet photons in the region.

Some YSOs are close enough that with Hubble or ground-based interferometers, we can resolve the disk without external illumination as is found in Orion. Edge-on disks provide particularly dramatic views of the disk, where a dark lane slices through the image of the YSO. The initial distribution of dust within a disk mimics the shape of a saucer: flat near the central star, and 'flaring' at larger distances. This characteristic shape reflects the combined effects of gravity pulling material downward toward the midplane, and thermal pressure of heated gas and dust pushing against the force of gravity.

■ **Left:** *An abundance of jets are revealed in this Spitzer Space Telescope image of a reflection nebula, NGC 1333, located 1,000 light years from Earth. The yellow-green knots of material throughout the image are glowing shock fronts from the jets, shot from young embryonic stars, as they plow into the surrounding nebula. Image courtesy NASA/JPL-Caltech/ R. A. Gutermuth (Harvard-Smithsonian CfA).*

■ *This Hubble Space Telescope view of a small portion of the Orion Nebula shows five young stars (the point-like objects), one showing a disk in silhouette and others shining by ultraviolet light. Image courtesy C.R. O'Dell/Rice University, NASA.*

In more distant star-forming regions, or in regions which lack the background illumination provided by the Orion Nebula, the presence of emerging YSOs can be inferred from Spitzer observations of spectral signatures specifically diagnostic of systems characterized by orbiting dust, illuminated and heated by a central star. The variation in brightness as a function of wavelength for a plain, disk-free star (one without dust around it) results in most of the light being produced at shorter wavelengths. In the case of a YSO with a disk of dust (and gas) around it, the dust is heated by the star and then produces its own infrared light, which changes the shape of the spectrum. There is then more infrared light than would be expected from a plain star, and this infrared excess betrays the presence of a disk.

The dust closest to the star is also the hottest, so its absence means that there is less near-infrared emission than from a complete disk. In these cases, the only dust producing infrared

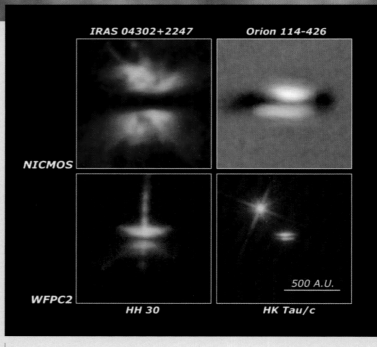

light is further away from the star, and much cooler, emitting only at mid- and far-infrared wavelengths. This resulting 'bump' or inflection in the spectrum indicates a disk with a missing center, and may be the first clue that planets have formed inside the disk.

■ *Above right: Images of edge-on disks observed by the Hubble Space Telescope. All images are shown to the same linear scale, with each box being about 20 times the diameter of Neptune's orbit. Images courtesy NASA and, left to right, top to bottom: D. Padgett (IPAC/Caltech), M. McCaughrean (Astrophysikalisches Institut Potsdam), C. Burrows (STScI), and K. Stapelfeldt (JPL/Caltech).*

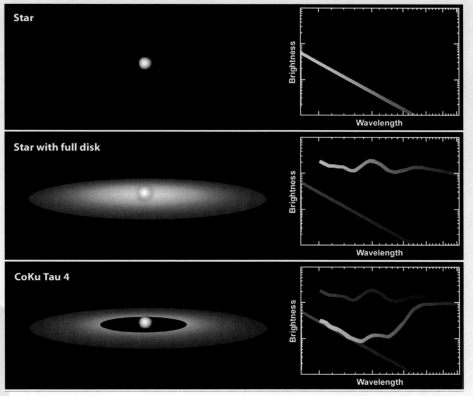

Using these patterns of brightness, Spitzer observations, combined with optical observations made from the ground, can provide a census of the number of forming stars that in fact are surrounded by disks. Studies of tens of star-forming regions, including the greater Orion complex, reveal thousands of likely YSOs, and find that more than 80 percent and probably all stars less than about a million years old – ranging in mass from 20 solar masses or more, to objects of mass far smaller than that needed to produce stars fueled by nuclear fusion – are surrounded by disks.

By using Spitzer's power to carry out a virtually complete census of star-disk systems (in relative isolation or in dense clusters) in star-forming clouds, it is possible to begin to study under what conditions stars form in relative isolation or in dense clusters, whether the formation of particular kinds of stars can trigger the formation of new generations of stars, and ultimately, how the stars that now populate the sea of stars in our Milky Way galaxies came to be over time.

The stellar nursery that pervades much of the constellation of Orion provides a clear demonstration of the diversity of star-formation outcomes in different clouds. The Orion A cloud is sprinkled with hundreds of isolated star-disk systems and a few small aggregates comprising 10-20 stars. In contrast, the Orion B cloud contains relatively few isolated stars, but several dense 'groups' of stars. Between the two clouds lies a region rich in newly-formed stars – the Orion nebula cluster. This cluster, home to the well-known Orion nebula, is destined in 100 million years or so to evolve into a cluster similar to the Pleiades. The stars in Orion A and B are unlikely to remain together for more than a few tens of millions of years after the gas and dust comprising the two clouds dissipates

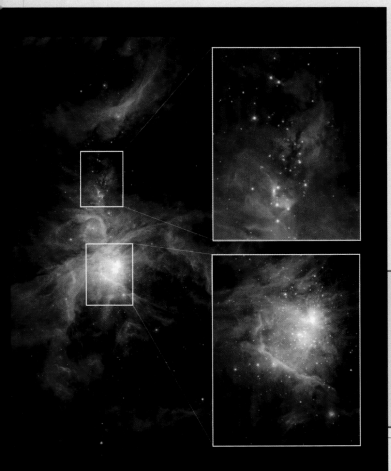

■ **Above:** *This diagram shows how we infer that a star has a protoplanetary disk around it, when the disk is too small to image directly. The left-hand column indicates the physical situation, and the right-hand column indicates the patterns of brightness as a function of wavelength (longer wavelengths on the right) that are observed. Images courtesy NASA/JPL-Caltech/D. Watson (University of Rochester).*

■ **Left:** *Spitzer observations of the Great Nebula in Orion (M42). Within the greater Orion complex, there are about 2500 young stars, including at least 200 very young stars. Images courtesy NASA/JPL-Caltech/ T. Megeath (University of Toledo).*

Infrared

Visible

This composite image shows an infrared image taken by NASA's Spitzer Space Telescope and a visible-light picture of the same region (inset, lower left). The so-called 'Mountains of Creation' reveal towering pillars of dust aglow with the infrared light of newly-forming embryonic stars (white/yellow). In visible-light the pillars are barely visible. The Spitzer image enables the dark clouds to be unveiled, revealing the process of star formation. Image courtesy NASA/JPL-Caltech/L. Allen (Harvard-Smithsonian CfA).

and releases these stars to join the billions of individual and isolated stars rotating around the center of the Milky Way.

What causes some clouds to favor more isolated star-forming events and others to form rich, dense clusters is currently unknown. Moreover, we don't know whether the mix of stellar masses found in dense clusters differs from that characterizing stars formed in isolation. A combination of Spitzer observations (which provide a census of newly-formed stars of all masses), radio observations (which can decode the physical, chemical and dynamical state of the cloud), and optical observations designed to estimate stellar masses will ultimately provide the clues needed to solve these mysteries.

More than 25 years ago, astronomers noted that regions that form stars with masses well in excess of 10 solar masses – regions like the Orion nebula – not only formed signature ionized hydrogen regions (gas fluorescing in response to the strong ultraviolet light emanating from massive stars), but also seemed to be associated with other groups of nearby, newly-formed stars. At the time, a few prescient researchers suggested that the formation of a massive star and its associated nebula of ionized gas could ultimately trigger the formation of another generation of young stars.

Spitzer's ability to pick out newly-formed star-disk systems via their signature colors has provided dramatic proof that triggered star formation occurs, and indeed may play a major role in turning clouds of gas and dust into stars. Here's how we think the process occurs. A newly-formed massive star creates an associated ionized hydrogen region and a plethora of newly-formed stars located at the rim defining the interface between the ionized region and the remainder of the cloud. The propagation of the hot, ionized material into the cold gas and dust in the surrounding star apparently compresses gas and dust at the boundary, forcing clumps of material into volumes small enough so that gravity can overcome internal pressure, thus beginning the collapse of a YSO.

The details of the triggering mechanism, and how the mix of stellar masses formed in triggered regions might differ from the mix found in regions which form stars 'spontaneously,' is yet another mystery that promises to find its ultimate solution through a combination of Spitzer and other observations.

Spitzer's census of star-disk systems can provide significant clues regarding the kinds of solar systems that can form around stars like the Sun (as well as stars of other masses), and ultimately answer a key question: is a Solar System like ours a typical outcome of the star formation process, or is it rare? Follow-up observations with powerful millimeter-wave telescopes capable of estimating the total disk mass from observations of heated grains suggest that at least half of the disks detected by Spitzer have mass sufficient to form a planetary system similar to our own. More modest systems – perhaps lacking our gas giant planets – could well form around the remaining 50 percent. Hence, most, if not all stars form with disks of mass sufficient to form a planetary system, and at least half can in principle assemble planetary systems similar to our own (as judged by total available disk mass).

Following the formation of a star-disk system isolated from its parent protostellar core, the micron-size dust embedded within the largely gaseous disk begins to settle toward the midplane – much as dust raised by a windstorm eventually settles earthward, clearing the air. Dust settling to the midplane results in an increase in the density of solid material, a fact that leads to very frequent collisions among the micron-size grains. Such collisions can quickly lead to merging (or fusing) of the micron-size grains into larger entities. Simulations – both in the computer and in the laboratory – suggest that once micron-size grains settle to the disk midplane, they can quickly form larger grains, and soon thereafter, kilometer-size bodies (within several hundred thousand years).

Above: *Artist's impression illustrating how planetary systems arise out of massive collisions between rocky bodies. In this image, a young star is shown circled by protoplanets and rings of dust. These rings – also called 'debris disks' – arise when protoplanets collide. One of these collisions is illustrated in the three inset panels above. Images courtesy NASA/JPL-Caltech/T. Pyle (SSC-Caltech).*

Collisions among these larger bodies, called planetesimals, can in turn produce even larger entities – planetary cores of lunar, Earth, and even tens of Earth-mass size. These collisions are accompanied by a 'second generation' of dust that within a few orbits gets smeared out into arcs or rings. In the inner regions of disks, near the orbit of the Earth, buildup from grains to Earth-size bodies can take place on timescales of several million years. Terrestrial planets (such as Mercury, Venus, Earth and Mars) are likely to form following this general picture.

However, forming gas giants similar to Jupiter and Saturn may require another process: the instigation of a cosmic 'snowstorm.' This happens when water vapor in the disk condenses to form icy grains which rain down to the disk midplane, increasing the density of solid material, and thus accelerating the formation of bodies of several earth masses in size. Such bodies are massive enough to gravitationally attract nearby orbiting gas and dust, rapidly building up a massive atmosphere around a solid core. Thus, Jupiter – the dominant gas giant in our own Solar System – may owe its origin to its unique location, provided, that is, that there is enough material in the disk to accumulate around the solid core to form a giant planet of Jovian (or supra-Jovian) mass.

HD69830 System

Earth's System

catastrophic effects on already-born terrestrial planets; the powerful gravitational field of the inward moving Jupiter could well eject these planets from their planetary systems. It may be that our Jupiter formed at just the right time – when there was enough material left in the disk to build a planet of its mass, and late enough, so that material in the outer disk did not force it to migrate inward.

In the outermost regions of the disk, the material thins out, and even in the dust-rich disk midplane is only able to form bodies of much lower mass. These outer disk regions may be the birthplaces of objects such as Pluto, and much smaller icy comets which populate the region exterior to Neptune's orbit in our own Solar System: the Kuiper Belt.

Eventually, the gas and dust in the disk is either assembled into planets or smaller bodies spanning a range of masses and orbital distances, accumulated by the central star, or gravitationally ejected from the forming system via gravitational interactions with the more massive orbiting bodies. The basic elements of other planetary systems are likely in place by no more than 10-30 million years after the initial core began its collapse.

Evolution of the planetary system continues, even billions of years after formation, as gravitational interaction among planets, comets, and asteroids results in readjustments of orbits, and ongoing cosmic collisions, which produce, as a by-product, micron-size dust grains. Our own Zodiacal light, visible with the naked eye at dawn and dusk (from a very dark location on a moonless night), provides vivid evidence of

Whether or not Jupiters in other systems are able to form, and remain orbiting their parent suns near the 'snow line,' depends critically on when they form and how much remaining disk material lies outside their orbits. Extrasolar Jupiters that form early, and in systems with massive outer disks, will, via gravitational interaction with outer disk material, be forced to migrate inward – perhaps explaining the presence of the large number of extrasolar Jovian-mass planets found orbiting much closer to their parent stars than our own Jupiter. Migration of such Jupiters from their place of origin may have

■ **Above:** *Artist's impressions of the light that comes from remnant disks (called the Zodiacal light) from a very dusty system, HD 69830 (upper panel), and, for comparison, from our own Solar System (lower panel). Images courtesy NASA/JPL-Caltech/R. Hurt (SSC).*

■ **Right:** *Spectra (middle four lines) from Spitzer of dusty disks around four brown dwarfs, in comparison to interstellar dust (top) and Comet Hale-Bopp (bottom). The light green vertical bands highlight the spectral fingerprints of crystals made up primarily of a green silicate mineral found on Earth called olivine. The broadening of these spectral features or bumps indicates silicate grains of increasing size. Image courtesy NASA/JPL-Caltech/D. Apai (University of Arizona).*

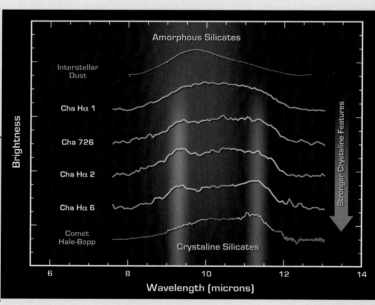

these ongoing processes in the light scattered Earthward by small, collision-produced dust grains. The Zodiacal light in our system is rather weak compared to some of the other, much dustier systems that astronomers have found (such as HD 69830). Evidence of collisions is found throughout the Solar System, most vividly etched in craters on the face of the Moon and other bodies lacking atmospheres. The collision-produced grains are heated to temperatures ranging from a cooked 1,000 degrees near the Sun to a chilly few tens of degrees near the outer reaches of our Solar System. These heated grains (here and in other planetary systems) can be detected by Spitzer, and provide a powerful probe of ongoing collisions among smaller bodies even in relatively mature planetary systems.

Stars much less massive than our Sun are likely to form and evolve the same way, only on longer time scales. Spitzer results show grain settling in disks around brown dwarfs, just as in more massive stars – comets have crystalline dust, and the interstellar dust does not.

At the other end of the scale, stars much more massive than our Sun have enough core temperature and pressure to ignite hydrogen very early in the process described above, quickly blowing away their natal material. Not only will those stars start hydrogen-burning early on, but eventually they are likely to explode as supernovae and trigger star formation in the surrounding molecular cloud, starting the cycle again.

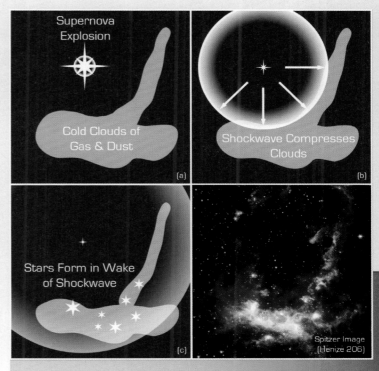

■ *Above: This four-panel diagram shows the process of triggered star formation. In the first panel (upper left), a massive, dying star explodes or 'goes supernova.' In the second panel (upper right), the shock wave from this explosion passes through clouds of gas and dust (shown in green). In the third panel (lower left), a new wave of stars is born within the cloud, induced by the shock from the supernova blast. The whole progression, from the death of one star to the birth of others, takes millions of years to complete. The fourth panel (lower right) is a Spitzer image of a region of the Large Magellanic Cloud called Henize 206, where it is thought this scheme is occurring. Images courtesy NASA/JPL-Caltech/R. Hurt (SSC-Caltech) and V. Gorjian (JPL).*

Dr Luisa Rebull is a staff scientist for the Spitzer Space Telescope, and specializes in star formation, in particular the evolution of rotation rates of young stars in the Orion Nebula. She earned her graduate degree in astronomy and astrophysics from the University of Chicago in 2000, and gives regular public talks in addition to her research.

Image courtesy Cynthia Lunine, 2000

Dr Stephen Strom has held appointments at Smithsonian Astrophysical Observatory and the State University of New York at Stony Brook, and was Chair of the Galactic and Extragalactic program at Kitt Peak National Observatory in Arizona. Following a 15-year appointment with the University of Massachusetts in Amherst, Strom returned to Arizona, joining the National Optical Astronomy Observatory where, despite recently retiring, he still carries out research into the formation of stars and planetary systems.

6

Chips off the PLANETARY BLOCK...

Star formation and planetary disk formation go hand in hand.
Here, astronomers *James R. Graham* and *Paul Kalas* both from the
University of California, Berkeley, explore the fascinating details astronomers
are now discovering about the formation of debris disks that linger after the

...Building Planetary
DEBRIS DISKS

The Zodiacal Light: *Our Own Debris Disk*

IN OUR own Solar System primitive bodies, such as comets and asteroids, are continually being eroded, releasing puffs and tails of debris, seeding interplanetary space with tiny dust grains. Born within the thin plane containing most of the solid mass orbiting the Sun, the trillions of dust grains form a tenuous, flattened cloud called a circumstellar disk. If gravity were the only force acting on these grains, then they would continue to orbit the Sun. However, various small but persistent influences, including radiation pressure from the Sun and drag forces, either cause grains to spiral inward to the Sun or blow them out to interstellar space. No worries. Asteroids and comets continue to fill the depleted debris disk with fresh grains. From our perspective on Earth, this dust is visible as the faint band of zodiacal light or the 'false dawn.' This is sunlight reflected from transitory dust in the plane of the Solar System.

The zodiacal light has been known since antiquity, but its counterpart for other stars was first seen in the early 1980s. Early clues to the existence of extrasolar dusty debris disks were found by George Aumann using the Infra-Red Astronomical Satellite (IRAS), which discovered that a handful of familiar, nearby A-type stars (stars with surface temperatures of 10,000 K), such as Vega and Fomalhaut, had orbiting dust clouds.

Seeing the zodiacal light is difficult – it can only be viewed on the darkest, moonless nights far away from the light pollution of cities. Imagine how hard it is too see the zodiacal light around other stars! In the 1930s French astronomer Bernard Lyot invented a special type of instrument for telescopes called a coronagraph to produce artificial eclipses of the Sun to permit routine observations of the tenuous solar corona. Shortly after the IRAS discovery, Brad Smith and Richard Terille used a coronagraph to show that the dust around Beta Pictoris was confined to a thin disk, and clearly analogous to the zodiacal dust in our own Solar System.

Ultimately, IRAS found that about 15 percent of all A-type stars have debris disks. Therefore, if debris disks arise from the erosion of larger bodies this was among the first hints that planetary systems are common.

The liquid helium cryogen for the IRAS satellite ran out on 21 November 1983, after

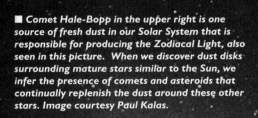

■ *Comet Hale-Bopp in the upper right is one source of fresh dust in our Solar System that is responsible for producing the Zodiacal Light, also seen in this picture. When we discover dust disks surrounding mature stars similar to the Sun, we infer the presence of comets and asteroids that continually replenish the dust around these other stars. Image courtesy Paul Kalas.*

nearly ten months of operation. However, the IRAS archival data have continued to be mined for neglected stars. As recently as 2004, a red dwarf sibling of Beta Pictoris, AU Microscopii, which was first detected by IRAS, was shown to have a prominent debris disk. The list of debris disks now includes many famous names, including Beta Pictoris, Vega, Fomalhaut, Epsilon Eridani, Tau Ceti, and Eta Corvi. Many more are now known, but are referred to merely by their entries in the 19th century star catalog of Henry Draper.

Debris Disk Sleuth:
The Spitzer Space Telescope

Although the IRAS catalog was the 'mother lode' for debris disk studies, radio telescopes operating at sub-millimeter wavelengths, the Infrared Space Observatory and the Hubble Space Telescope (HST) have all played a role in discovery and exploration. Since 2003, NASA's infrared Spitzer Space Telescope has surveyed hundreds of stars with unprecedented sensitivity. Like IRAS, Spitzer is effective at detecting warm dust heated by the luminous A-type stars – about one quarter now have detections – but Spitzer is for the first time able to find debris disks orbiting solar-type stars. For stars older than a billion years, cold Kuiper belt-like disks are not uncommon, but warm asteroidal belt dust still eludes detection.

Princeton astronomer Amaya Moro-Martin has been using Spitzer to study the inconspicuous 6th magnitude star HD 38529. She has shown that this system provides a remarkable example of the relationship between debris disks and planetary systems. HD 38529 is a several billion-year old G-type star 42 parsecs from the Sun. This is the only known dusty, mature twin planet system. HD 38529 hosts two eccentric planets: a close-in 0.8 Jupiter-mass planet at 0.13 AU and a 12 Jupiter-mass planet at 3.7 AU. In this case the dust is cold and likely arises from a broad belt ten times more distant from the parent star than its outermost planet.

Another striking Spitzer discovery is the intense, asteroidal-belt emission from the nearby red (K-type) dwarf, HD 69830, which hosts three Neptune-mass planets with short orbital periods that place them all within 0.63 AU. The dust emission from HD 69830 is one thousand times more intense than that from of our own zodiacal cloud and exhibits the infrared fingerprint of small, crystalline silicate grains located within 1 AU of the star. Intense asteroidal belt emission is rare (less than 2 percent) leading to speculation that this dust cannot be persistent, but is likely the consequence of the recent catastrophic disruption of an asteroid or large comet. Perhaps the giant impact that formed the Earth's Moon about four billion years ago produced a similar cloud of debris.

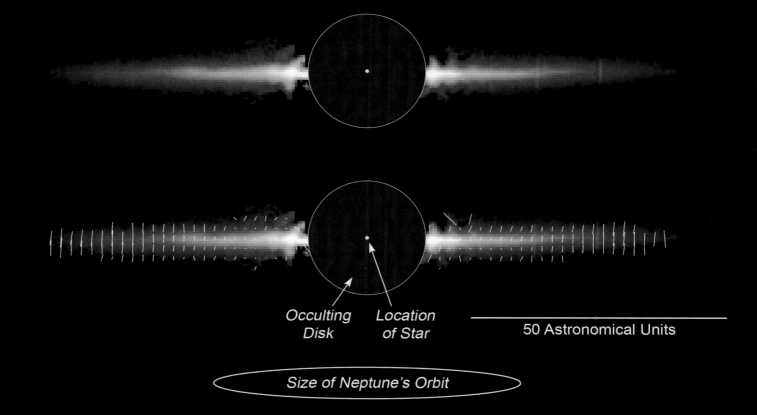

Occulting
Disk

Location
of Star

50 Astronomical Units

Size of Neptune's Orbit

Debris Disks and Planets

Why should we care about debris disks now that extrasolar planets can be detected, and over two hundred are known? One important reason is that the principal exoplanet detection technique, the Doppler method, only works for billion-year-old stars, similar to the Sun, that have stable outer layers and are spinning relatively slowly. Not only that, the Doppler method is essentially sensitive to extrasolar giant planets in relatively close-in orbits, missing the smaller objects similar to Uranus or Neptune in mass and distance from the star. The inner edges of debris disks, on the other hand, are sensitive to gravitational perturbations from anything that is like an extrasolar Neptune.

■ **Above:** *AU Microscopii, the red dwarf star and sibling of Beta Pictoris, has a debris disk that was first imaged in 2004. These recent Hubble pictures show that the system resembles an edge-on view of Saturn's rings. The dark circle corresponds to the coronagraph spot that blots out the light from the star. The lower panel shows the polarized light signature, revealing the fluffy or porous structure of the dust present in this system. Image courtesy James R. Graham, Paul G. Kalas, and Brenda Matthews, STScI/ESA/NASA.*

■ **Left:** *An artist's impression of the debris disk sleuth extraordinaire the NASA/Spitzer Space Telescope. This observatory is sensitive to the infrared wavelengths of 3 to 180 microns radiated by circumstellar dust grains. The observatory was launched from Cape Canaveral, Florida on 25 August 2003. The 85-cm telescope is cooled to below 6 degrees above absolute zero by 360 liters of liquid helium. Image courtesy NASA/JPL-Caltech.*

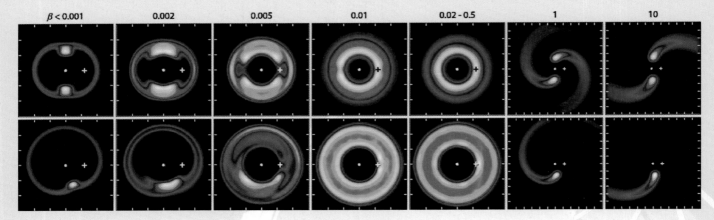

Over forty years ago, Tommy Gold, famous for developing the theory of steady state cosmology with Hermann Bondi and Fred Hoyle, realized that dust grains in our Solar System are likely to be driven into orbits that resonate with planetary orbits. If the gravitational pull of a planet like Jupiter were the only factor, then grains would be scattered out of the Solar System. In the 1950s Jan Oort and Gerard Kuiper envisioned that scattering by Jupiter and Neptune would populate the outermost regions of our Solar System with comets. This would also happen to dust grains, except that they are so tiny their motion is buffered by collisions with gas atoms, and their dynamics are further influenced by the scattering and absorption of sunlight. When combined with gravitational perturbations of planet, this portfolio of forces provides the dissipation necessary to concentrate matter into special orbits.

The relative importance of these forces depends on the size of grains, which means that grains are sorted according to how big they are. The largest grains have the same clumpy resonant distribution as comets and asteroids; moderate-sized grains are no longer in resonance and have an axi-symmetric distribution; and small grains are blown out of the system by radiation pressure and so have a density distribution that falls off inversely with distance from the star. As different wavebands – optical, infrared, and millimeter – are each sensitive to different dust grain sizes, a debris disk may have a very different appearance depending on the wavelength of light to which a camera is sensitive.

For example, take Vega's debris disk, where the entire system is thought to be oriented face-on from our point of view. If the Vega system were similar to the ringed planet Saturn, then our view is from the poles, and the rings of dust would neatly circle the star. Instead, when the 15-meter James Clerk Maxwell Telescope on Mauna Kea, Hawaii targeted Vega with its sub-millimeter

■ **Above:** *Computer models of dust grains orbiting an A-type star. Each panel shows the distribution of dust grains of different sizes: smallest grains are on the right; largest on the left. The grains are created in the collisional destruction of planetesimals trapped in the 3:2 resonance with a massive planet that migrated outward from 45 AU to 60 AU. The color scale represents the abundance of dust. The smallest grains (two panels on the right) feel the effects of radiation pressure and fly out of the system on hyperbolic orbits. The planet's location is shown with a white plus sign, and its orbital motion is counterclockwise. Image courtesy M.C. Wyatt, Royal Observatory Edinburgh, copyright The Astrophysical Journal, 639: 1153-1165, 2006 March 10.*

■ **Right:** *Two disks for the price of one! Very recent observations of Beta Pictoris with the Hubble Space Telescope show that a second disk lies hidden within the main disk. An occulting mask blots out the direct light from the star. Image courtesy D. Golimowski and NASA/ Space Telescope Science Institute.*

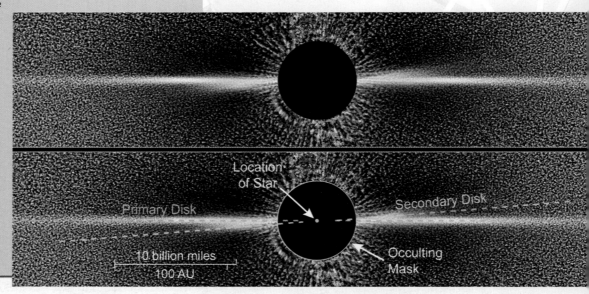

Location of Star

Primary Disk

Secondary Disk

Occulting Mask

10 billion miles
100 AU

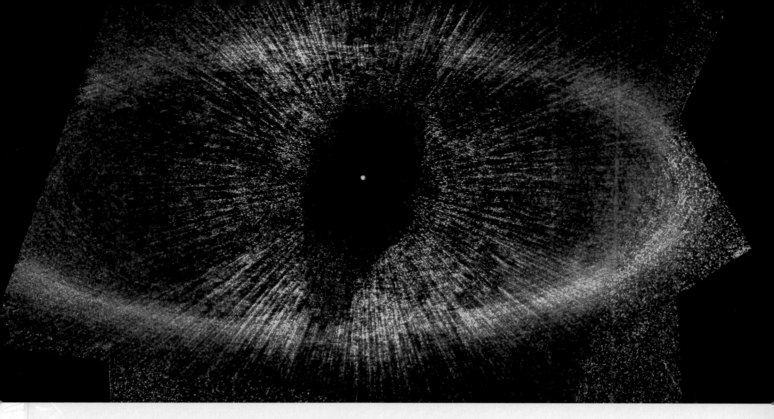

camera called SCUBA, the dust thermal emission was only found in one area 60 AU to the northeast of the star. Similarly, no neat circle of dust emission was found for Epsilon Eridani's dust disk – just a few clumps that if connected would form a ring 60 AU in radius. Fortunately, there is a very plausible explanation, which is that a yet undetected planet-mass object perturbs the dust into the resonances. Depending on the type of resonance (fundamental or harmonic), a star may have one or more concentrations of dust in its debris disk.

More Clues

For the Beta Pictoris and Fomalhaut debris disks it's not clumps that give away the presence of planets, but even more exotic features. Super-sensitive optical images of both stars using the Advanced Camera for Surveys aboard HST revealed that Beta Pictoris has a double disk, and that Fomalhaut is not a disk at all, but a narrow belt whose center is not at the position of the star, but 15 AU away from the star. The main Beta Pictoris disk, which has been studied in several hundred scientific papers over the past 25 years, is accompanied by faint sliver of light oriented a few degrees away from the main disk. Several theorists have proposed that an inner planet has an orbital plane that does not align with the main disk plane. The planet perturbs a belt of asteroids or comets vertically away from the main disk mid-plane, and dust from these primitive bodies

then gets shot out to greater distances, explaining this surprising phenomenon.

Fomalhaut, the bright southern star visible to the naked eye beginning in the late summer, has a beautiful ring of material inclined about 30 degrees from edge-on. If you trace the outline of the belt and find its geometric center, you would miss the star by 15 AU. At first glance the reader may recall Kepler's Laws of planetary motion, which state that orbits are elliptical, and that the star is found at the focus, not the center, of the ellipse. This is true for a single orbit, but if you have billions of orbits for a population of comets, the ensemble would appear as a circular ring centered on the star. It was only recently that theorists Mark Wyatt and Stan Dermott showed that if a planet has an eccentric orbit around the star, then it could align the periastra of all the other bodies next to it such that the ring of comets shifts its center away from the star. Thus, when we look at the offset of Fomalhaut's ring relative to the position of the star, we are witnessing the effect of an unseen planet whose orbit around Fomalhaut is not perfectly circular.

■ *Above: The most detailed visible-light image ever taken of a narrow, dusty ring around the nearby star Fomalhaut offers the strongest evidence yet that an unruly and unseen planet may be gravitationally tugging on the ring. The light from Fomalhaut itself is hidden behind an occulting spot in the camera. The radial spokes are image artifacts due to imperfections in the mirrors of the Hubble Space Telescope. Image courtesy Paul Kalas, James R. Graham, and Mark Clampin, NASA/Space Telescope Science Institute.*

Exotica

Frank Shu, the famous Berkeley astronomer and co-inventor of the density wave theory of spiral galaxies, once quipped that there were only two types of astronomical objects: spheres and disks. Astronomers have known about the spheres – stars and planets – for millennia. Only recently has it become apparent how common and important disks are. Ordinary stars are orbited by gas and dust at various stages of their formation and evolution, but even in their final configurations as white dwarfs and neutron stars do disks occur.

In the 1990s UCLA astronomers Ben Zuckerman and Eric Becklin were searching for brown dwarfs orbiting white dwarfs when they discovered that the white dwarf G 29-38 had an infrared excess emission. They attributed this anomaly to a cool stellar companion. One of us (JRG) showed that the emission was due to dust orbiting in an asteroidal belt close to the star. This model has stood the test of time and additional white dwarfs with debris disks, such as GD 362, have been discovered in the past few years. In 2006, Boris Gaensicke (University of Warwick, England) and colleagues showed that a disk of calcium vapor orbits the white dwarf SDSS 1228+1040, making a stunning confirmation of the white dwarf/disk picture. The characteristic double-horned Doppler signature of the spectral lines of ionized calcium means that the outer edge of the disk extends to only 1.2 solar radii. The likely origin of the disk is a tidally disrupted asteroid, which had been perturbed from its initial orbit by a relatively massive planetesimal object or even a fully-fledged planet.

It is often argued that the first extrasolar planets were found in 1992 orbiting the spinning, magnetized neutron star PSR B1257+12 using the 300-meter Arecibo radio telescope. These earth-mass 'pulsar planets' must have formed from the debris of the supernova explosion that formed the neutron star. The fact that the pulsar planets orbit in a single plane suggests that they condensed out of a disk. In 2006, Spitzer observed that a different pulsar 4U 0142+61 appears to be orbited by about ten earth masses of debris in a disk.

■ *Opposite page:* *Successful testing of two of the 12-meter telescopes for Atacama Large Millimeter Array at the Very Large Array in New Mexico took place in March 2007. Eventually about fifty telescopes will work together on the Chajnantor plain of the Chilean Andes, 5000 meters above sea level to form the world's most sensitive millimeter observatory. Image courtesy Drew Medlin, NRAO/AUI/NSF.*

The Future

While some facilities such as HST and Spitzer are approaching the end of their missions, other powerful new observatories are coming on line. The prime camera on Hubble for imaging debris disks – the Advanced Camera for Surveys – suffered a fatal short circuit on 27 January 2007. Spitzer's days are numbered as its cryogenic helium is slated to run out in April 2009.

Just about every large, ground-based observatory is planning major upgrades that will produce the sharpest possible images using adaptive optics, and suppress the light from disk candidate stars using coronagraphs and other innovative techniques. Before the end of the decade both the European Very Large Telescope (SPHERE) and the Gemini Observatory's planet imager (GPI) will deploy powerful adaptive optics systems that will see faint debris disks and their planets.

The Herschel spacecraft will launch in 2008, in a dual configuration with the European Space Agency's cosmic microwave background mission, Planck. The Herschel is equipped with a 3.5-meter telescope, which dwarfs the 85-cm Spitzer primary mirror. The instruments aboard Herschel will provide superb resolution at wavelengths as long as 200 μm and be complementary to the long anticipated SCUBA-II camera for the James Clerk Maxwell telescope.

In the high Atacama Desert of northern Chile a revolutionary new astronomical facility is taking shape on the Chajnantor plain of the Chilean Andes, 5000 meters above sea level. The Atacama Large Millimeter/submillimeter Array (ALMA) will be a radio interferometer, like the Very Large Array in New Mexico, comprising about 50 individual radio dishes. ALMA will be able to detect dust masses as small as one percent of the mass of the Moon and at the highest frequencies will achieve a factor of ten better angular resolution than the Hubble Space Telescope.

The 6.5-meter diameter James Webb Space Telescope, which is expected to launch in 2013, will have three times the resolution of HST and a suite of infrared instruments capable of imaging new debris disks with unprecedented sensitivity. If the last few years provide any guide, be prepared for more surprises about planets and planet formation as these unique new observatories uncover the remarkable world of debris disks!

James R. Graham is a professor of astronomy at the University of California, Berkeley, where he is project scientist for the Gemini Planet Imager project - an "extreme" adaptive optics system designed to allow direct detection of exoplanets. Previously, Graham was a senior research fellow at the California Institute of Technology, Pasadena. His PhD is from Imperial College, University of London.

Paul Kalas is an observational astronomer at the University of California at Berkeley. His research programs focus on high-contrast, high-resolution imaging of dusty disks around nearby stars. He has utilized some of the world's largest telescopes, and among his accomplishments are the optical discoveries of debris belts surrounding the nearby stars Fomalhaut and AU Microscopii using the Hubble Space Telescope. Kalas' coronagraphic observations of nearby stars represent the largest database of its kind.

7

High-Energy
VISION...

One of NASA's Great Observatories, the Chandra X-ray Observatory, is approaching a decade in orbit. Reviewing a string of stunning discoveries, science spokesperson for Chandra, *Wallace Tucker*, offers his unique and expert insight into what high energy X-rays are telling us about the most violent events in the universe.

...The Chandra
X-RAY
OBSERVATORY

NASA'S CHANDRA X-ray Observatory, named after the Indian-American astronomer Subrahmanyan Chandrasekhar, was launched and deployed by the Space Shuttle Columbia on 23 July 1999. Chandra has an unrivaled ability to make high-resolution images and temperature maps of the most energetic regions of the universe. This update reviews some significant advances made possible by Chandra in understanding the life cycles of stars, the role of supermassive black holes in the evolution of galaxies, and the study of dark matter and dark energy.

Out of Destruction, Renewal

One of the central quests of astronomy is to observe and understand the drama of stellar evolution, from the formation of stars in dense clouds of dust and gas to their demise, either quietly as white dwarfs, or violently as supernovas. Chandra and other X-ray telescopes focus on the high-energy action of this drama – sudden outbursts on the turbulent surfaces of stars, gale-force outflows of gas from hot, luminous stars, and awesome shock waves generated by supernova explosions.

Surveys of clusters of young sun-like stars (1 to 10 million years old) have revealed that they produce violent X-ray outbursts, or flares, that are much more frequent and energetic than anything seen today from our 4.6 billion-year-old Sun. The range of flare energies is large, with some of the stars producing flares that are a hundred times larger than others. The extent to which this flaring activity affects the formation of planets and the subsequent possibility of life evolving there is not well understood.

According to some theoretical models, large flares could produce strong turbulence in a planet-forming disk around a young star. Such turbulence might affect the position of rocky, Earth-like planets as they form and prevent them from rapidly migrating toward the young star. Possibly, large flares from the young Sun could have enhanced the odds for the Earth forming and remaining in the Sun's "habitable zone."

While sun-like stars will shine for billions of years, massive stars lead short, spectacular lives. After only a few million years, a star that is a dozen or more times as massive as the Sun will be using energy prodigiously and rushing headlong toward a supernova catastrophe. First, the massive star will expand enormously to become a red giant, and eject its outer layers at a

■ **Left:** *Artist's impression of the Chandra X-ray Observatory in orbit. Chandra has an unusual, highly elliptical orbit that takes the spacecraft to an altitude of 133,000 km – more than a third of the distance to the Moon – before returning to its closest approach to the Earth of 16,000 kilometers. It takes approximately 64 hours and 18 minutes to complete an orbit. Illustration courtesy CXC/NGST.*

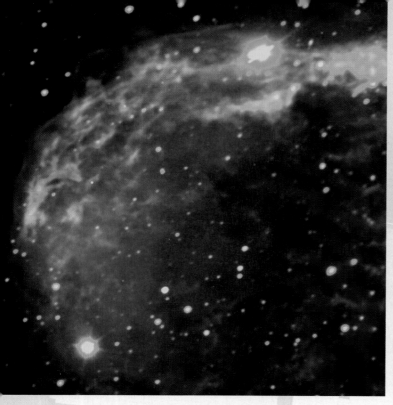

A few hundred thousand years after the onset of the fast stellar wind phase, a massive star is doomed. The nuclear energy supply in its interior can no longer hold up the weight of the overlying layers of the star, and the core collapses. Matter in the core is crushed to extreme densities. Electrons are pushed inside protons to form one of the strangest objects in nature – a neutron star, where a mass equal to the Sun has been crammed into a region no more than 20 kilometers across.

The formation of the neutron star releases enormous amounts of energy in the form of heat and neutrinos creating a titanic shock wave. This shock wave sweeps through the remaining outer layers of the star, fuses lighter elements into heavier ones and produces the brilliant optical display of a supernova, which shines with the power of a few hundred million suns for several months.

In the meantime, as the rapidly expanding (about 10 million kilometers per hour) stellar debris pushes into surrounding gas, two X-ray emitting shock waves are formed. One moves ahead of the debris, the other moves back into the outrushing debris.

After 10 or 20 years, the outward-moving shock wave hits the shell of gas produced by the slow and fast winds from the pre-supernova star. Just such an event is currently on display in Supernova 1987A, where optical and X-radiation from the shock-heated inner edge of the gaseous shell has become the predominant source of radiation from that object.

As the outer shock wave races past the shell into interstellar space, the inner shock heats the stellar debris to millions of degrees. These combined shock waves, which are tens of light-years across and glow in X-rays for thousands of years, produce supernova remnants, which provide some of Chandra's most striking images. The study of supernova remnants enables astronomers to the trace the progress of the shock waves and distribution of the elements ejected in the explosion – elements such as carbon, nitrogen, oxygen, silicon and iron that are necessary for planets and life.

Cassiopeia A (Cas A) is a spectacular example of a young (about 300 years old) supernova remnant. A recent, very long exposure of Cas A with Chandra reveals exquisite detail of the outer shock wave. An analysis of the data shows that charged particles are efficiently accelerated to extremely high energies there. The image also shows heated material from the inner shock wave.

speed of about 30,000 kilometers per hour. A few hundred thousand years later - a blink of the eye in the life of a sun-like star - the intense radiation from the exposed hot, inner layer of the massive star begins to push gas away at speeds in excess of 5 million kilometers per hour!

When this high speed 'stellar wind' rams into the slower red giant wind, a dense shell is formed. The force of the collision creates two shock waves: one that moves outward, lighting up the dense shell, and one that moves inward to produce a bubble of million-degree Celsius X-ray emitting gas.

Massive stars can lose half or more of their mass through stellar winds. The momentum from the radiation-driven winds creates large bubbles in surrounding clouds of dust and gas, which can trigger the formation of a new generation of stars. Observations of these hot bubbles by Chandra give new insight into an energetic phase in the evolution of massive stars.

■ *Above: Massive stars lead short, spectacular lives. This composite X-ray (blue)/optical (red and green) image reveals dramatic details of a portion of the Crescent Nebula, a giant gaseous shell created by powerful winds blowing from the massive star HD 192163 (a.k.a. WR 136, the star is out of the field of view to the lower right). After only 4.5 million years (one-thousandth the age of the Sun), HD 192163 began its headlong rush toward a supernova catastrophe. First it expanded enormously to become a red giant and ejected its outer layers at about 30,000 kilometers per hour. Two hundred thousand years later – a blink of an eye in the life of a normal star – the intense radiation from the exposed hot, inner layer of the star began pushing gas away at speeds in excess of 5 million kilometers per hour! Images courtesy NASA/UIUC/Y. Chu & R. Gruendl et al. (X-ray); and SDSU/MLO/Y. Chu et al. (Optical).*

■ *This composite image of Supernova 1987A shows the effects of a powerful shock wave moving away from the explosion. Bright spots of X-ray (blue-purple) and optical (pink-white) emission arise where the shock collides with structures in the surrounding gas. These structures were carved out by the wind from the destroyed star. Images courtesy NASA/CXC/PSU/S.Park and D.Burrows (X-ray); and NASA/STScI/CfA/P.Challis (Optical).*

In some supernova remnants, a rapidly rotating neutron star can produce a pulsating source of radiation (or pulsar) and a pulsar wind nebula of high-energy particles. The Crab Nebula, a remnant of a supernova observed in 1054 AD, is the most famous object of this kind.

Chandra has been able to detect numerous pulsars and their associated pulsar wind nebulas. These discoveries are proving to be one of the best ways to identify supernova remnants produced by the core collapse of a massive star and to distinguish such supernova remnants from remnants produced by the thermonuclear disruption of a white dwarf star.

A white dwarf star, the condensed remains of what used to be a sun-like star, is intrinsically the most stable of stars, as long as its mass remains below the so-called Chandrasekhar limit of 1.4 solar masses. However, if matter from a companion star is pulled onto the white dwarf, or if it merges with another white dwarf, it can be pushed over the Chandrasekhar limit. The white dwarf will then become unstable and be disrupted by a thermonuclear explosion.

An expanding cloud of ejecta glows brightly for many weeks as radioactive nickel produced in the supernova explosion decays into cobalt and then iron. These so-called Type Ia supernovas are thought to be the source of most of the iron in the universe. Analysis of a recent Chandra image of the supernova observed in 1604 AD, commonly referred to as Kepler's supernova, revealed a relatively high number of iron to oxygen atoms in the hot gas and no evidence for a neutron star. These results strongly support the identification of Kepler's supernova as a Type Ia supernova.

Out of Darkness, Light

Observations with Chandra and other telescopes imply that up to a quarter of the total radiation in the universe emitted since the Big Bang comes from material falling toward black holes, especially the supermassive black holes located in the central regions of galaxies. That's quite an accomplishment for objects that represent the universe's ultimate sinkhole from which nothing can escape. Recent observations with Chandra have shed light on the fascinating question of how black holes help to light up the universe.

The short answer has been known for years – the enormous gravity of a black hole pulls surrounding gas toward it and accelerates it to very high energies. As the gas forms in a disk and spirals toward the black hole, a portion of the energy of the gas is "somehow" expelled from the vicinity of the black hole before the matter passes beyond the event horizon – the edge of the cosmic sinkhole.

Computer simulations indicated that the "somehow" part of the explanation could involve magnetic fields embedded in the gas. Magnetic fields help to generate friction in a disk of gas swirling around a black hole. This friction could heat the gas, fueling the powerful light output from the disk. Magnetic fields also can serve to drive gas away from the black hole in winds and high-speed jets.

■ **Below:** *The supernova remnant Cassiopeia A (Cas A) was likely produced by an explosion of a massive star that had previously ejected most of its outer layers. In the Chandra image, blue, wispy arcs reveal particles accelerated to extremely high energies by an expanding shock wave generated by the explosion. More slowly moving shock waves in the interior have heated gas from the destroyed star to millions of degrees (red and green). The central white dot is believed to be a neutron star produced in the explosion. Image courtesy NASA/CXC/MIT/UMass Amherst/M.D.Stage et al.*

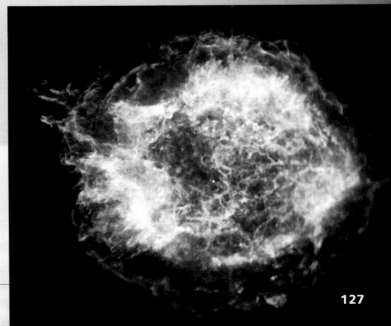

■ The Crab Nebula is the remnant of a supernova observed in 1054 AD. The bright nebula is due to the activity of a magnetized, rapidly rotating neutron star (bright white dot in the center of the image) that is spewing out a blizzard of extremely high-energy particles. Chandra's image reveals rings and jets of high-energy particles that appear to have been flung outward over great distances from the neutron star. The diameter of the inner ring is about 1,000 times the diameter of our Solar System. Image courtesy NASA/CXC/ASU/J.Hester et al.

■ Below: The Vela pulsar exhibits a similar ring and jet structure to that found around the Crab Nebula pulsar. Image courtesy NASA/CXC/PSU/G.Pavlov et al.

Recent Chandra observations of a wind from a disk around a black hole in our Galaxy have provided strong evidence for this general picture. The X-ray spectrum of a binary star system consisting of a black hole and a normal star indicates that much of the hot gas is spiraling inward toward the black hole, but about 30 percent is blowing away. The temperature and intensity of the winds imply that the only plausible method for producing these winds is through the action of powerful magnetic fields in the disk.

Somewhat paradoxically, the supermassive black hole in the center of our Galaxy appears to generating less commotion in the way of X-radiation, winds, or high-speed jets than many of its much less massive counterparts. Sagittarius A* (Sgr A*), as this supermassive black hole is known, is about 4 million times as massive as the sun, and is gaining weight daily as it pulls in more material.

The mystery is why Sgr A* is not growing faster. All the matter being spewed out by the many massive stars in the central regions of our Galaxy should provide Sgr A* with a good steady source of food. Yet the X-ray power of Sgr A*, normally a good indicator of the rate of mass being swallowed by a black hole, is unusually low. Chandra has caught Sgr A* in the act of snacking – it produced a series of bright flares – but the amount consumed was small, about the weight of a comet.

One explanation for Sgr A*'s severe diet is that the gas around Sgr A* is simply too hot, so we are seeing it in a quiet period. Another is that the winds from the massive stars in the vicinity are blowing too fast to be captured by the supermassive black hole. A black hole is in a way like a big, slow dog. If a rabbit stays far enough away, it can escape, but if it ventures too close…

Jets and other features observed in the central regions of our Galaxy suggest that Sgr A* was much more active in the past. Like cold case investigators, astronomers have used Chandra to uncover evidence of a powerful outburst generated in the past by gas falling toward Sgr A*. While some of the X-rays from the outburst would have traveled directly to Earth, X-rays that were reflected off gas clouds in the vicinity of the black hole took a longer path and arrived 50 years later, in time to be recorded by Chandra. Astronomers believe a mass equivalent to the planet Mercury was devoured by the black hole in this event.

Black Hole Blowback

On a galactic scale, images of galaxy clusters made by Chandra and radio telescopes have revealed evidence for the repetitive and far-

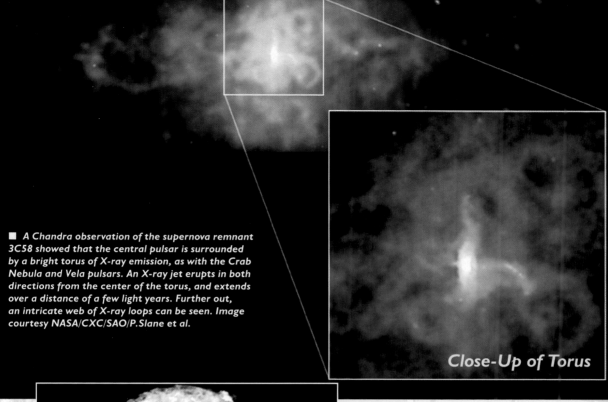

■ *A Chandra observation of the supernova remnant 3C58 showed that the central pulsar is surrounded by a bright torus of X-ray emission, as with the Crab Nebula and Vela pulsars. An X-ray jet erupts in both directions from the center of the torus, and extends over a distance of a few light years. Further out, an intricate web of X-ray loops can be seen. Image courtesy NASA/CXC/SAO/P.Slane et al.*

Close-Up of Torus

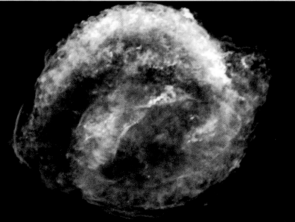

reaching explosive activity generated by supermassive black holes. Galaxy clusters are the largest gravitationally bound systems in the universe. A large cluster contains hundreds to thousands of galaxies immersed in an enormous cloud of multimillion degree hot gas and dark matter. The multimillion-degree gas clouds in clusters are strong X-ray sources that can extend over several million light-years.

A Chandra observation of the hot gas in the Perseus galaxy cluster shows wavelike features that appear to be sound waves. These waves are thought to be produced by the expulsion, every ten million years or so, of high-energy jets from the vicinity of a supermassive black hole in the huge galaxy at the center of the cluster. In another cluster, MS 0735.6+7421 (MS 0735, for short), Chandra found evidence for what may be the most energetic explosion ever detected, with

a total energy equivalent to the kinetic energy release of about 10 billion supernovae. Each of the two opposing X-ray bubbles spans a region more than 600,000 light-years across, roughly six times the diameter of the disk of our Galaxy.

The massive explosions observed in these and other clusters appear to be part of a cosmic feedback cycle in which the central black hole regulates its own growth, and the growth of the galaxy in which it resides. The inflowing gas fuels rapid growth of the supermassive black hole and produces high-energy jets and intense radiation. As energy flowing from the vicinity of the growing black hole ejects more and more gas from the galaxy, star formation rapidly slows, and the growth of the black hole declines.

This scenario is complicated and enhanced by galaxy collisions, an ever-present hazard in the central regions of galaxy clusters. Smaller galaxies passing too close to the giant central galaxy will be torn asunder by its gravitational tides. The stars of the smaller galaxy will be assimilated by the central galaxy, and some of the gas in the doomed galaxy will migrate inward to feed the supermassive black hole. The smaller

■ **Above left:** *Chandra's image of the remnant of Kepler's supernova observed in 1604 shows high-energy X-rays (blue) from extremely energetic electrons at the location of a rapidly moving outer shock wave generated by the explosion. The interior contains gas heated to millions of degrees by a more slowly moving shock wave. The composition of the hot gas and the absence of a neutron star indicate that Kepler's supernova was a Type Ia event. Image courtesy NASA/CXC/NCSU/S.Reynolds et al.*

galaxy's central black hole will eventually merge with the giant galaxy's supermassive black hole, increasing its mass. The enormous bubbles observed in the MS 0735 galaxy cluster were probably the result of an event such as this.

Dark Matter and Dark Energy

The hot gas in a galaxy cluster is in equilibrium with the gravitational pull of all the mass in the cluster. X-ray studies of numerous clusters have shown that 70 to 90 percent of the mass of a typical cluster consists of dark matter, which cannot be seen with any type of telescope and can be detected only through its gravity.

Numerous, indirect strands of evidence for the humbling statistic that most of the universe is made of stuff we can't see has been around for some time now, but more than a few astronomers have clung to the hope that a modification of the laws of gravity might eliminate the need for dark matter.

Those hopes were effectively dashed by a Chandra observation of a violent collision in the Bullet Cluster between two massive sub-clusters of galaxies. In this collision two huge gas clouds

exert a drag force, similar to air resistance, on each other. The dark matter, in contrast, slows down because it does not interact directly with itself or the gas except through gravity. This difference in behavior between normal and dark matter should cause them to separate. The stars behave like dark matter in the collision, because they do not collide, either.

Using data from the Hubble Space Telescope, the Magellan and ESO 2.2-meter telescope, and a technique known as gravitational lensing, astronomers showed that the mass concentration in the clusters is separate from that of the X-ray producing hot gas clouds observed by Chandra. This separation of normal and dark matter cannot be explained by the proposed modifications of gravity.

What could the dark matter be? One possibility suggested by superstring theory for elementary particles is the neutralino, which could have been produced in abundance in the first trillionth of a second of the Big Bang, and is predicted to have

■ **Above:** *This 400 by 900 light-year mosaic of several Chandra images of the central region of our Milky Way galaxy reveals hundreds of white dwarf stars, neutron stars, and black holes bathed in an incandescent fog of multi-million-degree gas. The supermassive black hole at the center of the Galaxy is located inside the bright white patch in the center of the image. The colors indicate X-ray energy bands - red (low), green (medium), and blue (high). The hot gas appears to be escaping from the Galactic Center into the rest of the Galaxy. The outflow of gas, chemically enriched from the frequent destruction of stars, will distribute these elements into the galactic suburbs. Image courtesy NASA/UMass/D. Wang et al.*

■ **Right:** *This composite optical (white), radio (red) and X-ray (blue) image of galaxy cluster (MS0735) shows dozens of galaxies, diffuse 50 million-degree gas permeating the space between the galaxies, and jets of high energy particles ejected from a central supermassive black hole. Images courtesy NASA/CXC/Univ. Waterloo/B.McNamara (X-ray); NASA/ESA/STScI/Univ. Waterloo/B.McNamara (Optical); and NRAO/Ohio Univ./L.Birzan et al. (Radio).*

Chandra X-Ray (3 color)

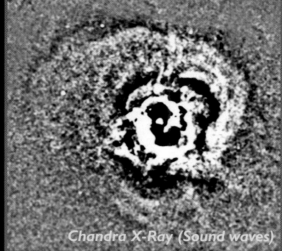
Chandra X-Ray (Sound waves)

■ *A specially processed Chandra observation of the central region of the Perseus galaxy cluster (left) has revealed wavelike features (right) that appear to be sound waves. These sound waves are thought to have been produced by explosive events occurring around a supermassive black hole (bright white spot) in Perseus A, the huge galaxy at the center of the cluster. Image courtesy NASA/CXC/IoA/A.Fabian et al.*

a mass about 100 times that of the proton. At the other end of the mass spectrum is the axion, which has a mass much smaller than a proton or even an electron. In between is a peculiar kind of massive neutrino, called a sterile neutrino.

None of these, nor any of the many other dark matter candidate particles, has ever been observed experimentally. But even the most elusive particle may leave a clue, and the ingenuity of astronomers and experimental physicists can be amazing, so someday soon, the nature of dark matter may come to light. However, solving the dark matter problem could be a piece of cake compared to achieving enlightenment on the nature of dark energy.

Astronomers have observed that the visible light from Type Ia supernovas, which have fairly well-known properties is fainter than expected in distant galaxies. The best explanation is that they are more distant than originally thought, which implies that the expansion of the universe must be accelerating.

Chandra's measurements of the dark matter content of clusters of galaxies have provided independent evidence for this astounding result. The data are consistent with a model in which the expansion of the Universe was first decelerating until about six billion years ago, and then began to accelerate.

Cosmic acceleration can be explained if the space between galaxies is filled with a mysterious dark energy. The existence of dark energy and its peculiar properties may require either a modification of Einstein's theory of general relativity or a major revision of some other area of fundamental physics.

■ **Above:** *This image of the galaxy cluster 1E0657-56 (a.k.a. the 'bullet cluster') shows hot X-ray producing gas (pink), and optical light from stars in the galaxies (orange and white). Using a technique known as gravitational lensing, astronomers have deduced that the mass concentration in the clusters (blue) is separate from that of the hot gas. This separation, which was produced by a high-speed collision between two clusters of galaxies, provides direct evidence that most of the matter in the clusters is dark matter. Images courtesy NASA/CXC/CfA/M.Markevitch et al. (X-ray); NASA/STScI; Magellan/U.Arizona/D.Clowe et al. (Optical); and NASA/STScI, ESO WFI, Magellan/U.Arizona/D.Clowe et al. (Lensing Map).*

Wallace Tucker, Science Spokesman for the Chandra X-ray Center, has authored or coauthored five popular books on astronomy. His most recent book, co-authored with Karen Tucker, is 'Revealing The Universe, The Making of the Chandra X-ray Observatory,' published in 2001 by Harvard University Press.

8

Surfing Long Waves
OF THE
UNIVERSE...

As America's National Radio Astronomy Observatory celebrates
more than 50 years of outstanding science, NRAO's Public
Information Officer, *Dave Finley*, brings an insider's look at the
stunning science from the past year and the remarkable facilities
planned for the future.

Image courtesy Tony Morris.

... The National Radio ASTRONOMY OBSERVATORY in the 21ST CENTURY

IN 2006, the National Radio Astronomy Observatory (NRAO) passed the 50-year point in its service to the astronomical community and now looks forward to an exciting future as one of the world's premier resources for answering the key questions that challenge 21st-century astronomy and physics. Today's world-class suite of NRAO telescopes provides unique and powerful capabilities for frontier research. Combined with effective user support programs, extensive archives of observational data, and expanding software packages, NRAO telescopes will continue to be among the most vital tools available to astronomers in the coming decades.

NRAO, a facility of the National Science Foundation, is busy, along with numerous international partners, building the Atacama Large Millimeter Array (ALMA), a groundbreaking project that will open a whole new window of 'discovery space' to the world's research community. ALMA will give unprecedented insights into the evolution of the early Universe and into the processes of star and planet formation. The Very Large Array (VLA), already the most versatile and productive ground-based telescope in the history of astronomy, is being expanded with state-of-the-art technology to keep it on the forefront of science.

■ **Left:** *The Atacama Large Millimeter/ submillimeter Array (ALMA), an international project now under construction in northern Chile, will provide unprecedented observing capability in a scientifically exciting range of wavelengths. This artist's impression shows some of the 12-meter-diameter antennas at the 5000-meter-high site. Image courtesy NRAO/AUI and computer graphics by ESO.*

ever-more-capable research tools that enabled landmark discoveries in an amazingly wide variety of astronomical specialties.

For example, NRAO telescopes were used to discover the black hole at the center of the Milky Way, to measure the bending of radio waves by the Sun to test the predictions of General Relativity, to discover the dark-matter haloes of galaxies, to measure the size of the afterglow of gamma-ray bursts, to make the most accurate distance measurement ever to a

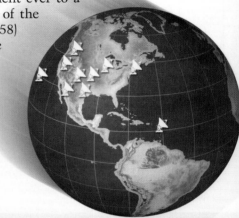

The Robert C. Byrd Green Bank Telescope (GBT), the world's largest fully-steerable radio telescope, continues to expand its impressive capabilities with new instrumentation. The continent-wide Very Long Baseline Array (VLBA) routinely offers the highest-resolution images of any astronomical instrument in the world.

In 1958, the NRAO, an organization then only two years old, dedicated its first telescope in Green Bank, West Virginia. That first instrument was followed in subsequent years with

galaxy not a satellite of the Milky Way (NGC4258) and thus recalibrate the distance scale of the Universe, to discover microquasars, and to discover dozens of molecules in interstellar space, to name just a few.

■ **Above:** The Very Large Array (VLA), on the high desert plains west of Socorro, New Mexico, is the most scientifically-productive ground-based telescope in history. It is now undergoing extensive improvements as part of the VLA Expansion Project, which will multiply its scientific capabilities by a factor of 10. Image courtesy NRAO/AUI and Kristal Armendariz.

■ **Above right:** The Very Long Baseline Array (VLBA) consists of ten 25-meter dish antennas distributed from Hawaii in the west to St. Croix, U.S. Virgin Islands, in the east. The VLBA provides the greatest resolving power (ability to see fine detail) of any telescope available, and has made landmark discoveries in several scientific areas. Image courtesy NRAO/AUI.

■ **Right:** The Robert C. Byrd Green Bank Telescope (GBT), with a diameter of 110 by 100 meters, is the world's largest fully-steerable antenna. Dedicated in 2000, it has become a leading tool for several important research areas, including the discovery of interstellar complex molecules and pulsar observations.
Image courtesy NRAO/AUI.

Now, with more than 9,000 refereed scientific papers and at least 575 Ph.D dissertations to its credit, the NRAO is focused on helping researchers answer the exciting and challenging questions that remain at the frontiers of science.

As astronomical research increasingly becomes a multi-wavelength enterprise, the NRAO's current and future suites of telescopes are ideally equipped to coordinate with instruments at other wavelengths. Observers routinely use NRAO radio telescopes in conjunction with optical, X-ray, infrared, and other facilities, including the Hubble Space Telescope, the Chandra X-ray Observatory and the Spitzer Space Telescope.

Versatility is a hallmark of NRAO's radio telescopes, which have been used to make major discoveries about objects ranging from the Sun and planets of our own Solar System to large-scale structures in the Universe and the nature of some of the first galaxies.

As an example of how NRAO radio telescopes can produce valuable new information that advances our understanding of important astronomical processes, let us examine some recent discoveries about how stars and planets are formed.

Stars and their planets form from giant interstellar clouds of gas and dust. Using the GBT, an international team of scientists has discovered new, complex, biologically-significant molecules in a pair of such clouds. In just two years the team, led by Jan M. Hollis of the NASA Goddard Space Flight Center, found eight new pre-biotic molecules.

The newly-discovered molecules all contain carbon and consist of between 6 to 11 atoms each. Found in a cloud called Sagittarius B2(N) near the Milky Way's center 26,000 light-years from Earth and in the relatively nearby Taurus Molecular Cloud (TMC-1) only 450 light-years away, the complex molecules suggest that the first of many chemical processes that ultimately led to life on Earth probably took place even before our planet was born.

The scientists discovered the molecules by using the GBT's great sensitivity to detect telltale emission or absorption at precise radio frequencies as the molecules in the gas

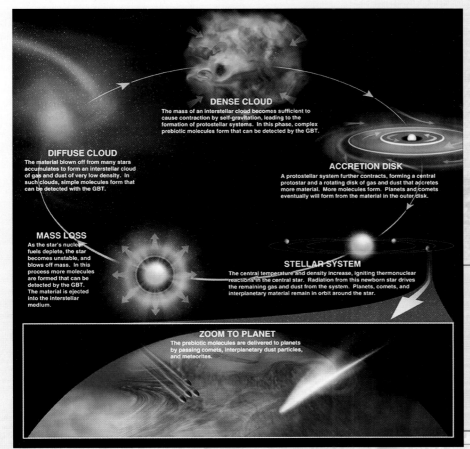

DENSE CLOUD
The mass of an interstellar cloud becomes sufficient to cause contraction by self-gravitation, leading to the formation of protostellar systems. In this phase, complex prebiotic molecules form that can be detected by the GBT.

DIFFUSE CLOUD
The material blown off from many stars accumulates to form an interstellar cloud of gas and dust of very low density. In such clouds, simple molecules form that can be detected with the GBT.

ACCRETION DISK
A protostellar system further contracts, forming a central protostar and a rotating disk of gas and dust that accretes more material. More molecules form. Planets and comets eventually will form from the material in the outer disk.

MASS LOSS
As the star's nuclear fuels deplete, the star becomes unstable, and blows off mass. In this process more molecules are formed that can be detected by the GBT. The material is ejected into the interstellar medium.

STELLAR SYSTEM
The central temperature and density increase, igniting thermonuclear reactions in the central star. Radiation from this newborn star drives the remaining gas and dust from the system. Planets, comets, and interplanetary material remain in orbit around the star.

ZOOM TO PLANET
The prebiotic molecules are delivered to planets by passing comets, interplanetary dust particles, and meteorites.

■ *Left: Stars and their planets form from giant interstellar clouds of gas and dust. Using the GBT, scientists have discovered new, complex, biologically-significant molecules in such clouds. These complex molecules suggest that the first of many chemical processes that ultimately led to life on Earth probably took place even before our planet was born. Illustration courtesy Bill Saxton, NRAO/AUI.*

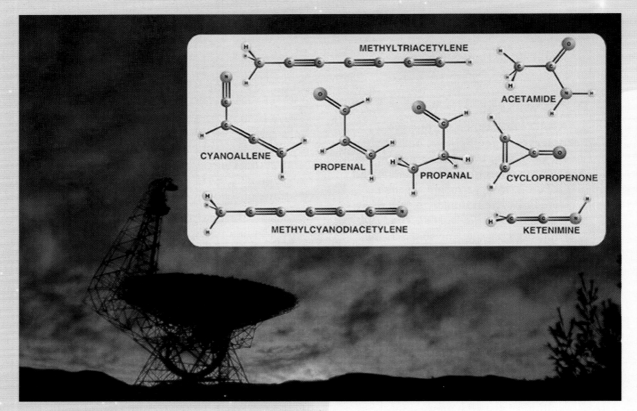

Above: *Large organic molecules form naturally within cold interstellar clouds through chemical processes that occur on the surface of dust grains. Shown here are a number of important molecules, precursors to amino acids and proteins, detected by the GBT, shown in the background. Image courtesy Bill Saxton, NRAO/AUI.*

clouds change their rotational modes. Laboratory experiments done before the astronomical observations showed which radio frequencies are characteristic of each of the molecules.

Finding relatively large organic molecules in cold regions of interstellar clouds led the astronomers to conclude that such molecules can be built up from smaller ones through chemical processes occurring on the surfaces of dust grains and in the interstellar gas itself. One of the molecules discovered with the GBT, acetamide (CH_3CONH_2), contains a peptide bond that links amino acids together to form proteins.

To improve our understanding of how stars form, astronomers need to determine the physical characteristics, such as mass and luminosity, of young stars in various stages of development with as much precision as possible. One barrier has been the imprecision of distance estimates to the few relatively nearby star-forming regions that have been the most extensively studied. That imprecision has limited the ability of observations to provide constraints on theoretical models of star formation.

The imprecision in distance estimates to star-forming regions arises because the visible light from young stars is dimmed by the cloud of gas and dust from which they are being formed. In addition, young stars also often are surrounded by material that obscures their visible light. As a result, even the Hipparcos satellite, which measured distances to more than 100,000 stars, was handicapped when it came to doing that for very young stars.

However, processes occurring in the magnetic fields of newly-formed stars often produce bright radio emission. Taking advantage of that, Laurent Loinard, of the National University of Mexico, led a team using NRAO's Very Long Baseline Array to make greatly improved distance measurements to nearby star-forming regions. The VLBA, a system of ten radio telescope antennas stretching from Hawaii to the Caribbean, provides the best ability to see fine detail, called resolving power, of any astronomical tool in the world. The VLBA can routinely produce images hundreds of times more detailed than those produced by the Hubble Space Telescope, reaching a resolution of less than half a milliarcsecond.

With the VLBA's tremendous resolving power, Loinard's team measured the tiny apparent shift of the star's position against the background sky caused by Earth's rotation around the Sun. This method, called trigonometric parallax, produces a direct, unambiguous measurement of the star's distance. The results were dramatic. For example, earlier work had placed a famous young stellar system in the constellation Taurus at a distance between 423 and 489 light-years. The VLBA measurements narrowed the range to 418-422 light years, reducing an uncertainty of 66 light-years to only four light-years.

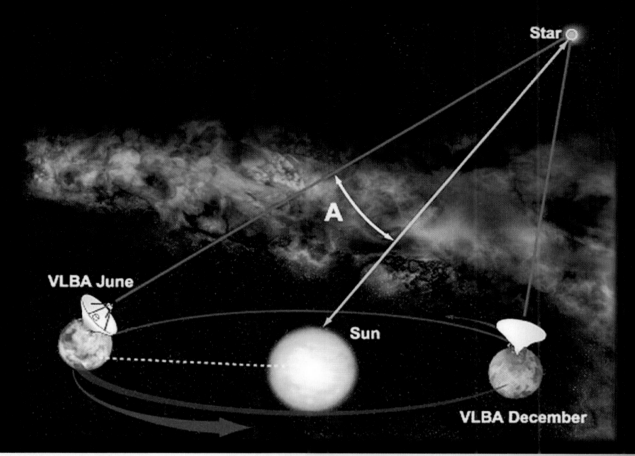

The improved distance measurements, which the team is extending to other star-forming regions, allow better determinations of the young stars' physical characteristics, such as mass and intrinsic luminosity. Also, by measuring the distances to multiple stars in a star-forming region, the scientists can build a three-dimensional picture of the region. Over time, the great resolving power of the VLBA will allow them to track the motion of individual stars within the region. Such data will provide valuable new insight into the processes of star formation.

Over the years, scientists have built a basic theoretical model for how Sun-like stars are formed, but have found it difficult to determine how very massive stars – those with ten times the

Sun's mass or more – can form. As gravitational collapse compacts material from the parent cloud of gas and dust onto the new star, theorists thought the process would naturally reach a mass limit. When a star reaches about eight times the Sun's mass, it pours out enough light and other radiation to stop further infall of material.

However, there are many stars more massive than that, so scientists were left with the question of just how they got that big.

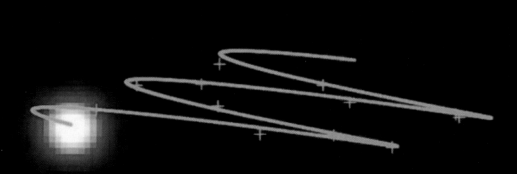

■ *The apparent path on the sky of the young star T Tauri S, caused by the Earth's motion around the Sun and measured with the VLBA. This measurement allows a direct determination of the distance to the star. Image courtesy Loinard and Mioduszewski, NRAO/AUI.*

To test this model, Maria Teresa Beltran, of the University of Barcelona in Spain and a team of collaborators studied a young, massive star called G24 A1, some 25,000 light-years from Earth. This object is about 20 times more massive than the Sun. Using the VLA, Beltran's team traced motions in gas around the young star by studying radio waves emitted by ammonia molecules. They measured the Doppler shift in the frequency of the radio waves to learn about the motions of the gas.

Their observations of G24 A1 were the first to detect all three types of motion predicted by the disk model – infall, outflow, and rotation – in a single massive young

One way around this problem is for infalling material to form a disk whirling around the star. With most of the radiation escaping without hitting the disk, material can continue to fall into the star from the disk. According to this model, some material will be flung outward along the rotation axis of the disk into powerful outflows. This model predicts that, as a massive young star is forming, there should be material falling inward, rushing outward, and rotating around the star all at the same time.

■ **Above:** *First stages of planet-building around a young star. This illustration shows pebble-sized chunks in the dusty disk surrounding the young star TW Hydrae. Astronomers believe such chunks are an early stage in a process that will build up larger-sized pieces until a planet ultimately is formed. Image courtesy Bill Saxton, NRAO/AUI.*

■ **Right:** *Young massive star system showing three types of motion. Artist's impression of the young stellar system G24 A1, showing material falling inward toward the star, rotating around the star, and being flung outward from the poles of the surrounding disk. Image courtesy Bill Saxton, NRAO/AUI.*

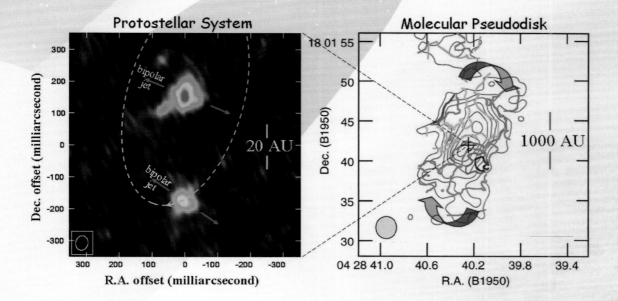

Protostellar System

Molecular Pseudodisk

bipolar jet

bipolar jet

20 AU

1000 AU

Dec. offset (milliarcsecond)

R.A. offset (milliarcsecond)

Dec. (B1950)

R.A. (B1950)

star. This milestone provides important evidence supporting the non-spherical accretion model for massive star formation, and make the postulated disks a plausible mechanism for building stars up to that object's size, 20 solar masses.

Where there are young stars, astronomers expect that there also are planets forming from the dusty disks surrounding the young stars. The dust in these disks, scientists believe, clumps together into larger and larger chunks until some of the chunks reach the size of planets. That means that pebbles and fist-sized pieces of rock in such a disk represent the early stages of planet construction.

When David Wilner of the Harvard-Smithsonian Center for Astrophysics and his colleagues used the VLA to study a nearby young star, they found this early planet-building process well underway. Their observations of TW Hydrae, a 10-million-year-old star 180 light-years from Earth, revealed that the protoplanetary disk surrounding that star contains regions of centimeter-sized pebbles extending outward at least a billion miles from the star.

Their VLA observations were the first to show this important early step on the path from dust to planets. Wilner's team benefited from the fact that TW Hydrae is relatively nearby and also the right age to be forming planets. The star's protoplanetary disk contains more than enough mass to form one or more Jupiter-sized planets.

A dusty disk like that in TW Hydrae tends to emit radio waves with wavelengths similar to the size of the particles in the disk. Other effects can mask this, however. In TW Hydrae, the astronomers explained, both the relatively

close distance of the system and the stage of the young star's evolution are just right to allow the relationship of particle size and wavelength to prevail. The scientists observed the young star's disk with the VLA at several centimeter-range wavelengths. They discovered strong emission at wavelengths of a few centimeters, which they interpreted as convincing evidence that particles of about the same size are present.

Most stars the size of our Sun or larger are not single, but rather are members of multiple-star systems. Astronomers have been divided on how multiple-star systems are formed, producing competing models. In one model, a pair of protostars, along with their surrounding dusty disks, fragment from a larger parent disk. Another model says stars form independently, then one captures the other into a mutual orbit.

Jeremy Lim, of the Institute of Astronomy & Astrophysics, Academia Sinica, in Taipei, Taiwan, and Shigehisa Takakuwa of the National Astronomical Observatory of Japan, studied a set of young stars called L1551 IRS5, 450 light-years from Earth in the constellation Taurus. Invisible to optical telescopes because of the gas and dust, this object was discovered in 1976 by astronomers using infrared telescopes.

■ *Above: VLA Images of the multiple star system and the larger, surrounding gas and dust disk. At left, the protostars are shown with the direction of their orbits indicated. At right, the larger disk is shown in a contour map image, with the direction of its rotation indicated by arrows. The molecular pseudo-disk image at right is adapted from Momose et al., Astrophysical Journal, 504, 314, 1998. See the artist's impression of this region, and associated web links for further reading, in the news item for 15 December in the chapter 'A Year in News and Pictures' elsewhere in this book. Image courtesy Lim & Takakuwa, NRAO/AUI/NSF.*

■ *The Atacama Large Millimeter Array, or ALMA, plans to deploy a group of 64 moveable radio telescope antennas, each 12-meters wide, that will work together to study the universe from a site in the foothills of Chile's Andes Mountains. The array will be able to span from 150 meters to 14 kilometers. A specialized computer, called a correlator, will combine the information received by the antennas, performing 16,000 million-million (1.6×10^{16}) operations per second. ALMA's location in the Atacama Desert is one of the highest, driest places on Earth, making it ideal for astronomical research at millimeter wavelengths, which are absorbed by atmospheric moisture. Completion is planned for 2012. Image courtesy NRAO/AUI and computer graphics by ESO.*

A VLA study in 1998 showed two young stars orbiting each other, each surrounded by a disk of dust that may, in time, congeal into a system of planets.

Using improved technical capabilities installed on the VLA since that observation, Lim and Takakuwa re-examined the system. They were able to image the system with unprecedented detail, and discovered a third, previously-unknown young star. Their new study showed that the disks of the two main protostars are aligned with each other and also with the larger, surrounding disk. They also found that the orbital motion of these two protostars resembles the rotation of the larger disk. These facts, they said, constitute a 'smoking gun' supporting the fragmentation model as a mechanism for producing multiple stars.

The newly-discovered third protostar, however, muddied the waters a bit. Its disk is misaligned with those of the other two. That could be explained either by its having formed independently and been captured, or by gravitational interactions with the other two. While the third protostar leaves open the door to the capture model, Lim and Takakuwa concluded that the L1551 IRS5 system provides strong support for the fragmentation model. They intend to study the system further.

Another tantalizing object indicates that multiple pre-stellar clouds may contribute to a single young star system, resulting in a system in which planets orbit the star in opposite directions. In our own Solar System, all the planets orbit the Sun in the same direction.

Anthony Remijan, of NRAO, and NASA-Goddard's Hollis found that the inner and outer portions of a disk surrounding a young star some 500 light-years away in the constellation Ophiuchus are orbiting in opposite directions. They used the VLA to measure Doppler shifts in radio waves emitted by silicon monoxide (SiO) molecules. To their surprise, they found that the SiO molecules close to the young star are orbiting in a direction opposite to that of other molecules farther from the star that had been studied earlier.

This is the first time that such counter-rotation has been seen in the disk around a young star. The astronomers speculate that the material in the disk may have come from two interstellar clouds that were rotating in opposite directions. The object is in a large star-forming region where chaotic motions and eddies in the gas and dust can produce smaller cloudlets that rotate in different directions.

Each part of the disk contains enough material to form planets. That means that the solar system that may develop around this star could have planets orbiting in different directions – a dramatically different situation from our own familiar family of planets.

Similar stories of NRAO's telescopes making major contributions to forefront astrophysical problems can be told about many other scientific specialties, including multi-wavelength studies of the Sun, supernovae and supernova remnants, pulsars and general relativity, galactic evolution, black holes, gamma-ray bursts, cosmology, and others. Over the years, researchers from many institutions have found ways to apply the capabilities of NRAO's telescopes to scientific problems that were not contemplated by the designers of those instruments.

The observatory's staff of scientists, engineers, and technicians, constantly strives to improve the telescopes' capabilities, and each NRAO instrument will, as they have in the past, grow in its usefulness to the research community. Two projects now are underway to make major impacts on the kinds of science that can be done. These are the VLA Expansion Project and the Atacama Large Millimeter/submillimeter Array (ALMA).

The VLA, dedicated in 1980, is in the midst of a major expansion that will make it roughly 10 times more capable as a scientific instrument in several respects. The Expanded VLA (EVLA), based on the 230-ton dish antennas and current infrastructure, will offer wider frequency coverage, greater sensitivity, better imaging capability, and a vastly improved analysis capability to the current instrument.

ALMA, now under construction on the Atacama Desert of northern Chile, is a partnership of North America, Europe, and Japan, in cooperation with the Republic of Chile. ALMA will provide the world's astronomers with their best-ever tool for exploring the Universe at millimeter and submillimeter wavelengths. ALMA will detect fainter objects and be able to produce much higher-quality images at these wavelengths than any previous telescope system. Scientists are eager to use this advanced capability to study the first stars and galaxies that

formed in the early Universe, and to learn vital new details about star and planet formation. ALMA will complement the capabilities of the next generation of orbiting observatories.

Many scientists eagerly plan to use the current and future facilities of the National Radio Astronomy Observatory to advance their knowledge in numerous astronomical specialties. In many cases, they have solid expectations for the kinds of new information that the growing capabilities of these scientific tools will provide, and they are excited about their prospects for discovery. However, the history of astronomy clearly indicates that the greatest excitement will come from those completely unexpected discoveries that will transform our understanding of the Universe.

For further information, consult these websites:

NRAO web site:
http://www.nrao.edu/

ALMA web site:
http://www.alma.nrao.edu/

Introduction to Radio Astronomy:
http://www.nrao.edu/whatisra/

Young stellar system in Taurus:
http://arxiv.org/abs/astro-ph/0501062

G24 A1:
http://arxiv.org/abs/astro-ph/0609789

TW Hydrae:
http://arxiv.org/abs/astro-ph/0506644

L1551 IRS5 system:
http://arxiv.org/abs/astro-ph/0512225

Dave Finley **has been Public Information Officer at the National Radio Astronomy Observatory since 1992. He is a former science/medicine editor for The Miami Herald and a past president of The Albuquerque Astronomical Society. He is co-organizer of the Enchanted Skies Star Party held annually in Socorro, New Mexico.**

9

Through the Eyes of the
SLOAN DIGITAL SKY SURVEY...

As an ambitious sky survey enters its second phase,
team astronomer *Timothy Beers* shares some of
the most exciting science to come out of the Sloan
Digital Sky Survey.

*Image courtesy Robert Lupton and the
Sloan Digital Sky Survey Consortium.*

...Probing Deeper
INTO THE
Milky Way
GALAXY

FOR THE past six years, astronomers working with the Sloan Digital Sky Survey (SDSS) have been collecting the world's largest database of calibrated photometry for galaxies and stars. This now comprises more than 200 million objects over 8,000 square degrees of the northern sky, as well as medium-resolution spectroscopy for over one million galaxies, quasars, and stars in the Milky Way. The original SDSS, now known as SDSS-I, formally came to an end in July 2005. Since that time this ever-growing collaboration, which now comprises some 25 universities and institutions worldwide, and involving more than 300 individual scientists, have continued their journey of discovery with the first extension of SDSS, known as SDSS-II. SDSS-II includes three separate mini-surveys – if one can imagine anything 'mini' about SDSS – which effectively split the available time on the ARC 2.5m telescope on Apache Point, New Mexico.

The SDSS-II extension, which started in July 2005 and runs through July 2008, includes:

(i) LEGACY, which is an effort to obtain the small amount of remaining photometry needed to complete the SDSS-I footprint on the northern sky, as well as to obtain a substantial amount of spectroscopy for the numerous quasars that were not observed during SDSS-I.

(ii) The SUPERNOVA survey, which makes use of repeat scans (taken roughly a few days apart over the course of several months during each fall observing season) of a 300 square degree strip of sky (2.5 degrees wide by 120 degrees long) located on the celestial equator. SUPERNOVA seeks to quickly identify possible Type Ia supernovae with redshifts in the range 0.1 < z < 0.3, a region of redshift space that previously was undersampled relative to more nearby or more distant supernovae.

(iii) SEGUE, the Sloan Extension for Galactic Understanding and Exploration, the first dedicated SDSS effort to collect spectroscopy for a large number (250,000) of stars in the disk and halo systems of the Milky Way. SEGUE is also collecting some additional 3,500 square degrees of imaging at lower galactic latitudes than obtained during SDSS-I, so that astronomers can better understand the important transition region between the disk(s) and halo components of the Milky Way.

■ **Left:** *The Apache Point Observatory in the Sacramento Mountains in the Lincoln National Forest, nearly 30 kilometers south of Cloudcroft, New Mexico (USA). The ARC 2.5-meter telescope used for the Sloan Digital Sky Survey is visible lower left. Image courtesy Apache Point Observatory.*

Getting to Know the Milky Way

The imaging data from SDSS-I, and the supplemental imaging from SEGUE, is providing an enormously rich view of the nature of the closest galaxy in the entire Universe to us, our own! The fundamental limitation of trying to study the large-scale structure of the Milky Way is that, unlike studies of more distant galaxies, our perspective of the Galaxy is one of 'inside looking out', rather than one of 'outside looking in'. As a result, one has to inspect large portions of the entire sky around our location in order to make quantified statements about its stellar structural components. In the past, this could only be accomplished with laborious photographic surveys, taken by specialized telescopes located in the Earth's northern and southern hemispheres. As valuable as such data was to obtain, it suffered from many shortfalls, not the least of which are the intrinsic variability, non-linear response, and poor sensitivity of photographic emulsions, and the limited number of wavelength regions that could be inspected. The ARC 2.5m telescope, with its wide, 3-degree field of view, its ability to rapidly scan and record digital data in five separate filters, designated 'ugriz' in the nomenclature of SDSS, and its very high efficiency, now allows for a new picture of the Milky Way to be drawn – and the details are fascinating!

Galactic Streams, Overdensities, and 'Hobbit Galaxies'

Certain components of the Galaxy have been known for many years. These include the disk systems, comprising the thin disk, of which our Sun is a member, and the thick disk, which is a lower density stellar component that includes stars up to several kpc above the thin disk; the central bulge of our Galaxy, located some 8 kpc from the Sun in the direction of the Sagittarius constellation; and the halo of our Galaxy, which includes stars and globular clusters out to at least 100 kpc away, as well as the still poorly understood

■ *Above right: Close-up view of the ARC 2.5-meter telescope at Apache Point in New Mexico. Image courtesy the Sloan Digital Sky Survey (SDSS-II).*

■ *Right: The specialized CCD array for the Sloan Digital Sky Survey. Each chip has a different specific color filter, allowing five "ugriz" colors to be obtained for each object photographed. Such colors allow approximate properties of each object to be determined, and 640 of special interest will be selected for more detailed spectra at a later date. Image courtesy the Sloan Digital Sky Survey (SDSS-II).*

The ARC 2.5m Telescope

The telescope's optical system is dominated by two reflecting mirrors, a primary of 2.5m dimension, and a specially shaped secondary mirror that enables the wide field of view of this remarkable tool. The focusing system includes two corrective lenses that minimize distortion, so that the images formed by the system remain in good focus across the entire field of view, which is equal to an area of sky of about 30 full Moons.

Constructing a Two-Dimensional Image of the Night Sky

The SDSS camera is one the most complex imaging devices yet built. It includes 30 silicon electronic light sensors called charge-coupled devices, or CCDs, that are each two inches square. Scientists encase each column of five devices in a vacuum-sealed chamber. In order to enhance sensitivity, liquid nitrogen cools each chamber to -80 degrees Celsius. Each CCD is made up of more than four million picture elements, which release electrons as light is absorbed. The electrons in turn are amplified into electronic signals that can be digitized, and recorded for later analysis. Each of the five rows of CCDs receives the light through a different colored filter, so each row records the brightness of objects in a different color. A typical night of observing generates up to 200 Gbytes of data.

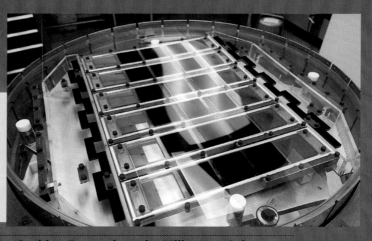

The SDSS camera assembles a two dimensional image by performing 'scans' across regions of sky, in such a way that the light from stars and galaxies traces across each of the columns, and in the process generates a series of images in the five 'ugriz' filters, but with a set of 'missing data' between each of the columns. This missing information is filled in by offsetting the direction that the telescope points by one half of the width of a CCD, and another scan is then made. The software onboard the camera is sufficiently smart that it can take out the 'smearing' of images that results from the movement of the sky across the detectors, so a 'static' image is obtained, similar to what would get by pointing in one direction at a time, but instead making it possible to record the desired information for swaths of sky that are many tens to hundreds of degrees in length, and 2.5 degrees in width. A simulation of this approach to strip mining the night sky can be found at: http://cas.sdss.org/dr5/en/tools/scroll/

Dispersing the Light from the Cosmos with the SDSS Spectrographs

Astronomers use a device called a spectrograph in order to spread the light received from certain pre-selected objects so that the spectrum can be recorded and later analyzed in order to learn numerous details about the nature of each celestial object. In order to obtain the spectra of many stars or galaxies and quasars at once, SDSS astronomers drill 640 holes into an aluminum plate, with each hole corresponding to the position of a selected object in the field of view of the ARC 2.5 m. Scientists then plug the holes, by hand, with optical fiber cables, which then simultaneously capture light from the 640 objects and record the results onto two large-format CCDs, each housed in its own spectrograph capable of receiving the light from 320 fibers. The plug plates are interchangeable with the CCD camera at the focal plane of the telescope. On a good night, observers will use six to nine such plug plates, and thereby obtain spectra of between 4000 and 6000 objects per night of operation.

dark matter that dominates the mass in the Galactic halo. In addition to these components, SDSS has recently demonstrated beyond all doubt that many new structures exist within it, or in the near vicinity.

Among the most interesting new structures revealed by SDSS are numerous Galactic streams, which are almost certainly the result of the destruction, or partial destruction, of low-mass smaller galaxies, often referred to as 'dwarf galaxies', by their interaction with the tidal field of the Milky Way during their orbits about the Galactic center. In some cases, the 'parent' of a stream may even have been a globular cluster that has been similarly ripped apart. Researchers in SDSS have begun to identify so many individual streams – over eight and counting at present – that one particularly striking image of the northern sky where several such examples are known is now referred to as the 'Field of Streams'. The most spectacular of these, the Sagittarius Stream, was known even before SDSS, but has been shown by studies with SDSS to clearly consist of multiple 'wraps' around the Milky Way, as the orbiting stars in this stream slowly dissolve into the general halo. Another fascinating example is the Monoceros Stream, discovered by SDSS researchers, that appears to be stellar debris left very close to the plane of our Galaxy from a recent interaction between a parent dwarf galaxy and the disk of the Milky Way.

Also present in the Field of Streams, and elsewhere in SDSS imaging, are examples of objects that do not appear to be coherent streams,

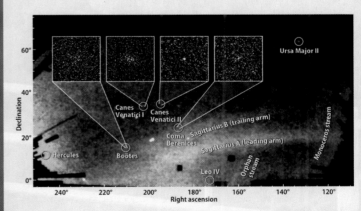

Above: The so-called "Field of Streams", assembled from SDSS-II imaging data, reveals streams of stars that are considered to be the result of the complete gravitational disruption of dwarf galaxies by tidal fields in the halo of the Galaxy, or following an attempted passage through the disk of the Milky Way. A number of newly identified low luminosity dwarf galaxies are also identified in this plot. Image courtesy V. Belokurov, the Sloan Digital Sky Survey (SDSS-II), and Astronomy Magazine.

Left: An astronomer inserts optical fibers into the many holes in the pre-drilled plug plate to allow multiple spectra of specific objects to be obtained. Image courtesy the Sloan Digital Sky Survey (SDSS-II).

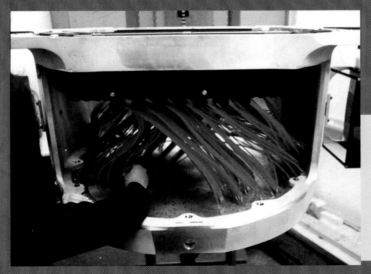

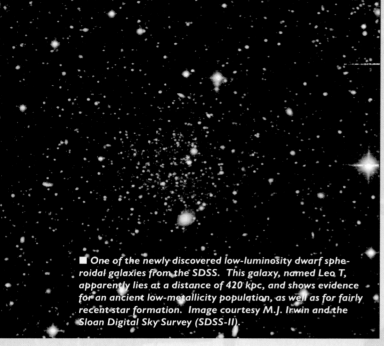

■ *One of the newly discovered low-luminosity dwarf sphe-roidal galaxies from the SDSS. This galaxy, named Leo T, apparently lies at a distance of 420 kpc, and shows evidence for an ancient low-metallicity population, as well as for fairly recent star formation. Image courtesy M.J. Irwin and the Sloan Digital Sky Survey (SDSS-II).*

numbers of stars (at most a few million), and are so spread out on the sky, typically covering regions of up to 30 arc minutes, roughly the size of the full Moon, that their dim collective glow completely fades away into the night sky. These nearly invisible galaxies have been possible to find with SDSS because it is able to resolve their individual stellar members, and separate them from the images of much more distant galaxies, which outnumber the stellar images at the low light levels reached by SDSS. The discovery of these eight new galaxies doubles the numbers of the low-luminosity dwarfs discovered in the region of the Milky Way over the past 70 years.

Vasily Belokurov, an SDSS researcher at the Institute of Astronomy in Cambridge, UK, where the group that has discovered most of these dwarf galaxies resides, has called them 'More like hobbits than dwarfs,' since they are so much smaller and fainter than most previously known satellites. Several of the newly discovered systems appear to be 'caught in the act' of their own destruction, and deep imaging has revealed that some, such as the Ursa Major II dwarf, already appear to be in several pieces. "They look as though they're being ground up," notes Belokurov.

The discovery of the 'hobbit galaxies' is of particular importance, since current theories of how galaxies form predict that many, perhaps all, of the stars in the halo and thick disk of the Milky Way originated in smaller dwarf galaxies, which were dissolved when they merged into the Milky Way itself. In fact, the current paradigm predicts that many more such faint dwarfs remain to be discovered. Keeping in mind that SDSS-I and SEGUE imaging presently only covers roughly one-fifth of the full sky, there should be at least a total of 40 additional examples found; in all likelihood many more will be found.

but rather, are enormous regions of the sky where there appear to be far more stars per unit area on the sky than would be expected from a smooth model of our Galaxy. Two examples discovered by SDSS are the Virgo Overdensity, and the Hercules-Aquila cloud. Although the final verdict remains uncertain at present, most astronomers assume that these overdensities may in fact be related to nearby dwarf galaxies, which appear so large on the sky because they are essentially 'in our face' – a classic case of not seeing the forest for the trees! One recent suggestion from SDSS scientist Martinez-Delgado and colleagues is that the Virgo Overdensity is in fact a large portion of the 'leading arm' of the Sagittarius Stream, which is presently falling on to the disk of the Milky Way. Additional studies with SDSS and other efforts will help reveal the true nature of these intriguing overdensities in the near future.

The SDSS can also lay claim to the discovery of some of the faintest, and perhaps most important galaxies, in the vicinity of the Milky Way. Astronomers have recognized, for many decades, the existence of a handful of higher luminosity, and hence relatively easy to identify, dwarf galaxies close to our Galaxy, or among the collection of galaxies – including our own, as well as the nearby Andromeda Galaxy – in local space known, appropriately, as 'The Local Group'. In the past two years, SDSS has identified eight new dwarf galaxies, objects that contain such small

■ **Right:** *The so-called Virgo Overdensity (visible as the red area near the top of the diagram) reveals a higher number of stars than would be expected from a smooth distribution in the halo of the Milky Way. It is not yet clear whether this overdense region is a relatively nearby portion of the "leading arm" of the Sagittarius Stream, which appears large because of its proximity to the Sun, or is in fact a very nearby dwarf galaxy in its own right. Image courtesy Z. Ivezic and the Sloan Digital Sky Survey (SDSS-II).*

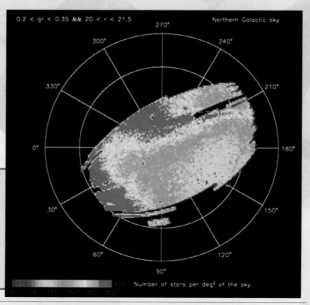

9 - Through the Eyes of the Sloan Digital Sky Survey . . . Probing Deeper into the Milky Way Galaxy

The Nature of the Stars and Stellar Populations in the Milky Way

As important as the extensive imaging of stars in the Milky Way is, the new information only scratches the surface of what astronomers hope to learn about its formation and evolution. More in-depth understanding requires more detailed observations, in this case from studies of the distribution of the constituent wavelengths of light from individual stars, the stellar spectra. Astronomers use the powerful SDSS spectrographs to simultaneously record the spectra of over 600 stars at a time; until the last ten years or so, technological limitations meant that similar information could only be obtained, laboriously, on a star-by-star basis. The full story of the nature of the various stellar components of the Milky Way is now being studied by this massive, 'industrial scale' effort.

During the operation of SDSS-I, numerous stellar spectra were obtained. Some of these were used for calibration of the galaxy and quasar spectra taken at the same time. Others, in particular for large samples of cooler, late-type M stars, and hotter, much more distant stars known as horizontal-branch stars, were selected because they were of special interest to several of the participating groups at the time. The real 'gold mine' for exploration of the stellar populations of the Galaxy is the 250,000 stellar spectra that are being obtained during the SEGUE campaign of SDSS-II. SDSS astronomers have used the colors of stars derived from the 'ugriz' imaging to select examples of a wide variety of objects that collectively provide the information needed to understand the nature of the stellar populations throughout the Galaxy.

Stellar Atmospheric Parameters, and the Jewels of the Night Sky

One key requirement for this effort is to be able to quickly, and accurately, measure the so-called atmospheric parameters of each star studied by SDSS-I and SEGUE. The outer atmospheres of stars absorb the light that comes from deep within each star in specific wavelengths, which atomic physics tells us corresponds to particular transitions that occur within the atoms of the stellar atmospheres, and from which the important parameters that can be used to describe the star are derived, in particular the stellar temperatures, surface gravities, and chemical compositions, all key ingredients used to assess the luminosity – and hence distance – and the origin of the star in the history of the Galaxy.

SDSS astronomers, led by the author, graduate student Young Sun Lee, and post-doctoral fellow Sivarani Thirupathi at Michigan State University, and involving others throughout the SDSS collaboration, have developed and refined a set of procedures known as the SDSS/SEGUE Stellar Parameter Pipeline (SSPP), from which we can derive this detailed information based on the appearance of the stellar spectrum and the colors of the stars. The SSPP is an assembly of a variety of techniques, some of which have been developed previously, others of which are custom made for analysis of SDSS spectra. These techniques are presently being validated and tested by comparison with objects of known temperatures, gravities, and composition, such as stars in globular clusters and individual stars for which much higher resolution spectra with other telescopes are being obtained. The SSPP is now able to measure the parameters of stars with the accuracy needed to conduct a host of detailed studies of individual stars in the Milky Way, an activity which is only just beginning in earnest.

One group of objects of great interest to SDSS astronomers are the stars which have chemical compositions that indicate they were born at the very earliest times, within several hundred million years following the Big Bang, when the Milky Way was not yet even a fully formed galaxy. SEGUE specifically selects, on the basis of colors, the stars which are likely to be among these prized 'jewels of the night sky'.

Astronomers sometimes refer to these stars as 'jewels' because of their extreme rarity. All of the elements heavier than hydrogen, helium, and a tiny amount of lithium, were formed as a result of nuclear processing in the hearts of stars, including many from the supernova explosions that occur when massive stars die. The early-generation stars that are being looked for by SEGUE are objects that were born from gas that was only 'polluted' by the (likely) high-mass stars that were the first objects formed in the early Universe. As other stars are born later, the gas from which subsequent generations are formed is polluted with an ever increasing amount of elements, such as the light elements carbon, nitrogen, oxygen, and metallic species such as calcium, magnesium, and iron. Thus, a star with very low abundances of such elements had to be formed at the very earliest stages of this process. In the local volume of the Galaxy near the Sun, the so-called very metal poor stars – those with abundances of elements less than one percent that found in the Sun – are identified with a frequency of only about one star in a million.

Inspection of spectra of stars from SDSS and SEGUE that are from a hundred to several thousand times more deficient in their metallic species than in our Sun reveals clear differences between them and solar-abundance stars; these data are unveiling a part of the cosmic history of the Milky Way. Astronomers have identified, with other survey efforts, roughly 2,000 very metal-poor stars over the past 50 years since they came to be recognized as a class. By the time SEGUE is completed in July 2008, there will be a total sample of some 20,000 such stars identified. Because of their importance for understanding the origin and evolution of the elements in the Galaxy and in the Universe, the very metal-poor stars identified with SEGUE will form the basis of high-resolution spectroscopic studies. From these, a much more detailed knowledge of their chemical content can be derived with the world's largest telescopes, such as the Subaru 8m, the Hobby-Eberly 9.2m, the VLT 8m, and Keck 10m, on up to the 30m and larger telescopes now being planned for construction in the coming decade. (For more on this topic, see the chapter entitled 'The History and Future of Telescopes...400 Years from Galileo to Keck and Beyond' by Jerry Nelson elsewhere in this volume.)

■ *Below: These spectra from SDSS and SEGUE show, from left to right, decreasing metallicity as measured by the ratio of the abundance of iron (Fe) to hydrogen (H) ([Fe/H]). The stars are a hundred to several thousand times more deficient in their metallic species than in our Sun. Among the 250,000 stars targeted for spectroscopy by SEGUE, there is the expectation of finding roughly 20,000 very low metallicity stars, from which a more detailed chemical history of stars in the Milky Way can be built. Image courtesy T.C. Beers and the Sloan Digital Sky Survey (SDSS-II).*

The Motions of Stars in the Milky Way

The spectra obtained of individual stars contain yet more information! In particular, the radial motion of a star, along the line of sight toward or away from us, causes a shift in the wavelengths of well-known absorption features in a star, a phenomenon known as a Doppler shift. The so-called 'radial velocities' that are recovered from measuring this shift tell astronomers that some stars are moving rather slowly, relative to the motion of the Sun, on the order of a few to tens of km/s. These stars are members of the thin-disk population, which includes the Sun. Other stars in the sky, all of which are still members of the Milky Way, exhibit radial velocities from many tens, up to over 500 km/s with respect to the Sun. These stars are part of either the thick-disk or the halo populations. Because of their large velocities, such stars move on orbits which can take them to the outermost regions of the halo of the Milky Way.

But wait, there's more! In addition, the precise positions of stars recorded with the SDSS cameras, when compared with previous photographic surveys taken many decades ago, can be used to measure the amount of angular motion over time of a star across the line of sight, that is, perpendicular to the radial direction. This motion – referred to by astronomers as the 'proper motion' – can be combined with the radial velocities, and the distances to the stars based on the parameters measured by the SSPP discussed above, to derive estimates of the full 'three dimensional motion' of the stars that are sufficiently close to subtend a significant angular displacement over time scales of several decades. In reality, the SDSS positions are used to re-calibrate, and thus greatly improve, the proper motions obtained from the previous surveys.

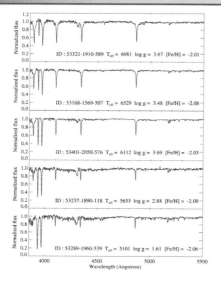
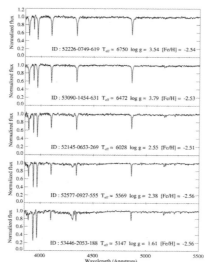
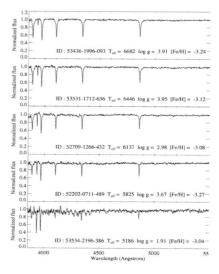

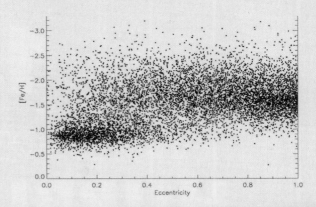

These data then allow astronomers to reconstruct the approximate orbits of the stars in a local volume, and thereby study the detailed kinematics of stars belonging to populations that occupy a much larger region of space.

One recent example of such studies comes from an analysis of the motions of stars used to calibrate the spectra of galaxies and quasars. These 'calibration stars' number over 25,000 objects, many of which are sufficiently close to have measurable proper motions, and hence full space motions may be derived.

A plot of the distribution of the metallicity of a subsample of these stars versus a derived orbital eccentricity, which is the shape of the orbits of the stars that can be inferred once one adopts a model for the distribution of mass throughout the Milky Way, shows that many stars with higher metallicities have quite low eccentricity (roughly circular) orbits, while many of the low-metallicity stars have very high eccentricity orbits. One also notices that there are large numbers of low-metallicity stars that exist with orbits at all eccentricities – this behavior is expected from models for the formation of the Galaxy from the destruction of numerous dwarf-like galaxies over the history of the Milky Way.

As reported by Daniela Carollo, a scientist at the Torino Observatory in Italy, and a recent visitor with my group at MSU, at the Seattle American Astronomical Society meeting in early 2007, this information is now revealing a very detailed picture of how the stellar populations in our Galaxy formed and evolved. Their study, and the exploration of the even larger sample of similar stars from SEGUE, is revolutionizing astronomers' understanding of the process of galaxy formation, a story that began over 12 billion years ago, and continues to the present.

Above: *The distribution of metallicity ([Fe/H]) for a subsample of stars in the solar neighborhood versus a derived orbital eccentricity, which is the shape of the orbits of the stars that can be inferred once one adopts a model for the distribution of mass throughout the Milky Way. The plot reveals that many stars with higher metallicities have quite low eccentricity (roughly circular) orbits, while many of the low-metallicity stars have very high eccentricity orbits. Image courtesy T.C. Beers, D. Carollo, and the Sloan Digital Sky Survey (SDSS-II).*

Funding for the SDSS and SDSS-II has been provided by the Alfred P. Sloan Foundation, the Participating Institutions, the National Science Foundation, the U.S. Department of Energy, the National Aeronautics and Space Administration, the Japanese Monbukagakusho, the Max Planck Society, and the Higher Education Funding Council for England.

The SDSS is managed by the Astrophysical Research Consortium for the Participating Institutions. The Participating Institutions are the American Museum of Natural History, Astrophysical Institute Potsdam, University of Basel, University of Cambridge, Case Western Reserve University, University of Chicago, Drexel University, Fermilab, the Institute for Advanced Study, the Japan Participation Group, Johns Hopkins University, the Joint Institute for Nuclear Astrophysics, the Kavli Institute for Particle Astrophysics and Cosmology, the Korean Scientist Group, the Chinese Academy of Sciences (LAMOST), Los Alamos National Laboratory, the Max-Planck-Institute for Astronomy (MPIA), the Max-Planck-Institute for Astrophysics (MPA), New Mexico State University, Ohio State University, University of Pittsburgh, University of Portsmouth, Princeton University, the United States Naval Observatory, and the University of Washington.

For further information, consult these websites:

The Sloan Digital Sky Survey:
http://www.sdss.org

About SEGUE:
http://www.sdss.org/segue/aboutsegue.html

The Field of Streams, Sagittarius and its Siblings, by Belokurov et al. (2006):
http://arxiv.org/abs/astro-ph/0605025

Discovery of an Unusual Dwarf Galaxy in the Outskirts of the Milky Way, by Irwin et al. (2007):
http://arxiv.org/abs/astro-ph/0701154

The Virgo Stellar Over-density: Mapping the Infall of the Sagittarius Tidal Stream onto the Milky Way Disk, by Martinez-Delgado et al. (2007):
http://arxiv.org/abs/astro-ph/0609104v2

The Joint Institute for Nuclear Astrophysics SEGUE Virtual Journal:
http://groups.nscl.msu.edu/jina/seguevj/

Timothy Beers *is University Distinguished Professor of Astronomy at Michigan State University. Beers received a PhD from Harvard University in 1983, and was a Bantrell Fellow at Caltech from 1983 to 1986 before joining the faculty at Michigan State. He has worked for almost a quarter of a century on searches for the earliest generations of stars in the Milky Way. Beers leads the SDSS-II team responsible for development of the stellar spectroscopic parameter pipeline for application to SEGUE.*

10
Reinventing the
HUBBLE SPACE TELESCOPE...

If all goes to plan, space shuttle astronauts will re-visit the Hubble Space Telescope in August 2008. Project scientist for the mission, *J. Chris Blades*, of the Space Telescope Science Institute, reviews the exciting array of science instruments to be installed.

...The Next
HUBBLE
Servicing Mission

SHUTTLE ASTRONAUTS will make one final 'house call' to the Hubble Space Telescope as part of a mission to extend and improve the observatory's capabilities through the year 2013. NASA Administrator Michael Griffin announced plans for a fifth servicing mission to Hubble on Tuesday, 31 October 2006. "We have conducted a detailed analysis of the performance and procedures necessary to carry out a successful Hubble repair mission over the course of the last three space shuttle missions. What we have learned has convinced us that we are able to conduct a safe and effective servicing mission to Hubble," Griffin said. "While there is an inherent risk in all spaceflight activities, the desire to preserve a truly international asset like the Hubble Space Telescope makes doing the mission the right course of action."

Mission planners have now settled on a launch-readiness date of 7 August 2008, for space shuttle Atlantis to carry out this important mission to Hubble, designated STS-125. The astronaut crew will include the veteran astronaut Scott D. Altman, commander; Gregory C. Johnson, pilot; with mission specialists including veteran space walkers John M. Grunsfeld and Michael J. Massimino, and first-time space fliers Andrew J. Feustel, Michael T. Good, and K. Megan McArthur. Scott Altman is no stranger to Hubble – he commanded Columbia in 2002 for STS-109, the last servicing mission, and John Grunsfeld has visited and worked on Hubble twice before. The crew is undergoing rigorous training for the mission, which includes training in the Neutral Buoyancy Laboratory where there is an HST mock-up.

■ **Left:** *The astronauts selected for SM4, the fifth and final shuttle mission to perform work on the Hubble Space Telescope, pose for a group photo. From left to right are astronauts K. Megan McArthur, Michael T. Good, Gregory C. Johnson (pilot), Scott D. Altman (commander), John M. Grunsfeld, Michael J. Massimino, and Andrew J. Feustel. Scott Altman commanded Space Shuttle Columbia in 2002 for SM3B, the last servicing mission, and John Grunsfeld has visited and worked on Hubble twice before. Image courtesy STScI/NASA.*

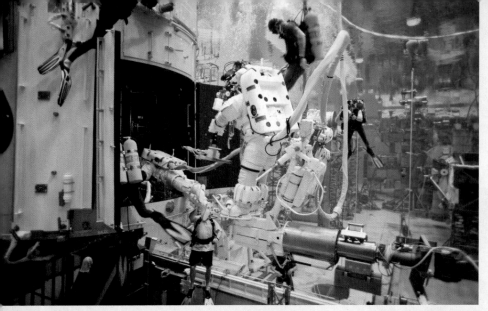

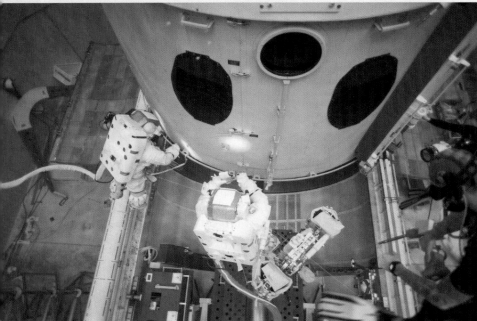

These new observations are changing our view of the Universe. Hubble has played a key role in studies of the mysterious form of energy called dark energy, which is acting like a cosmic gas pedal, accelerating the Universe's expansion rate. Observations of Cepheid variable stars in the Virgo cluster and other clusters have helped to determine a precise age for the Universe of 13 to 14 billion years old. Host galaxies of quasars have been clearly imaged, many for the first time; and Hubble studies of black holes have shown them to be ever present in the centers of galaxies, where they grow in size with their host, feasting on the galaxy's gas and stars (see article 'As NASA's Flagship Great Observatory Completes 15 Years in Orbit ... What's New with Hubble?' by Ray Villard in last year's volume, *State of the Universe 2007*).

It's worth remembering that Hubble is a cooperative project of two space agencies, NASA and ESA (the European Space Agency). Located about 600 kilometers above the Earth's surface and orbiting every 97 minutes with a path inclined 28.5 degrees to the equator, Hubble's altitude is near the limit for shuttle accessibility. Hubble's primary mirror, a modest 2.4 meters in diameter, allows imaging and spectroscopy over the wavelength region from the ultraviolet (~120 nanometers) to the near-infrared (2.5 microns), light that is normally filtered away by Earth's atmosphere. Its exquisite and stable imaging at a wide range of wavelengths is what makes Hubble such a powerful and unique astrophysical tool.

Ever since its launch in 1990, the Hubble has been in the public eye. Dogged initially by blurred vision from spherical aberration in the primary mirror, as well as a series of operational problems, a repair mission launched by NASA in 1993 transformed the observatory into the highly successful mission it has since become. A steady stream of out-of-this-world pictures has captured the public's imagination, while astronomers worldwide continue to be excited by the huge amount of astrophysical data returned by Hubble.

Servicing Mission 4 (SM4) will actually be the fifth visit to Hubble since the telescope's launch in April 1990. The repair mission, SM1, took place in December 1993, SM2 in February 1997, SM3A in December 1999, and SM3B in March 2002. (SM3 was split due to a critical need to replace gyroscopes in 1999.) SM4 was originally planned for 2004, but the February 2003 Columbia tragedy led to its postponement

■ *Above: The Servicing Mission 4 astronauts will practice many hours in preparation for the upcoming mission to Hubble. Some of this time will be spent under water going through mission simulations at NASA Johnson Space Center's Neutral Buoyancy Laboratory (NBL). Top: With equipment bay doors open, a crew member prepares to work inside of Hubble. Bottom: Two astronauts maneuver around the outside of Hubble. Images courtesy NASA Johnson Space Center and STScI.*

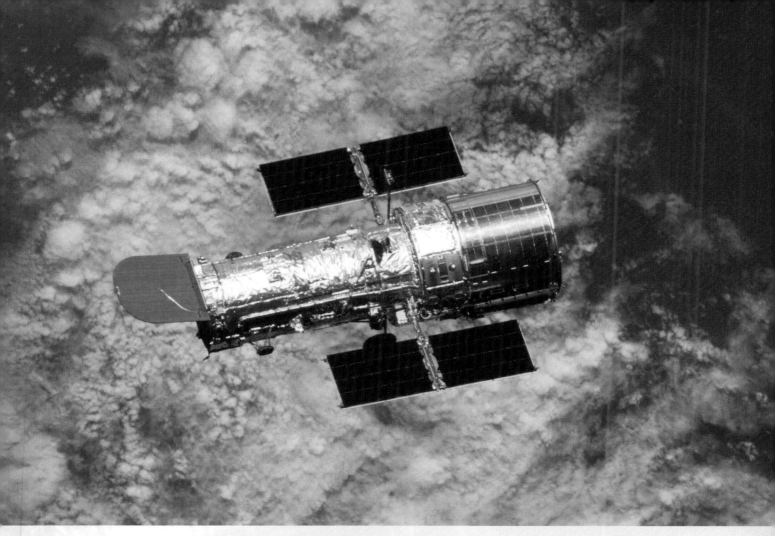

and eventual cancellation due to safety concerns. Following three successful flights, significant improvements in the shuttle, and a re-examination of servicing mission risks, NASA considers it safe to fly the shuttle back to Hubble.

Since SM3B, the observatory has lost use of two major science instruments and is operating with viewing restrictions because of further gyroscope failures. Today, astronomical images are being obtained only with the venerable Wide Field Planetary Camera-2 (WFPC2, installed in 1993), and in the near infrared with NICMOS (the Near Infrared and Multi-object Spectrograph), which is the only axial instrument in use. Even though it is now suffering from radiation damage, the WFPC2, stoutly designed by the Jet Propulsion Laboratory, has proven to be a life saver for Hubble. The Space Telescope Imaging Spectrograph (STIS) that was installed during SM2 provided a wealth of unique spectroscopic results in the ultraviolet and optical for a number of years. Unfortunately, an electrical short blew a fuse in May 2001, knocking out its primary electronics bank. A switch was made to the

redundant electronics (all instruments have two separate electronics sides), and the instrument returned to science operations until a power supply in the redundant side failed in August 2004, rendering STIS unavailable for science.

The Advanced Camera for Surveys (ACS) quickly took over the role of workhorse camera from WFPC2 following its installation in 2002 on SM3B. Its modern detectors and specially coated optics enabled a huge amount of cutting edge science, including some remarkable survey programs, such as the mapping of the Orion Nebula, a new Hubble Deep Field (the Ultra-Deep Field) and a wide-field cosmology survey code-named COSMOS (reviewed by Anton Koekemoer in last year's volume, State of the Universe 2007). Unfortunately, electronics failures in it too have rendered it largely lost, except for a little used ultraviolet camera. In

■ *Above: Floating against the background of Earth, the Hubble Space Telescope gets back to work after a week of repair and upgrade by space shuttle Columbia astronauts during SM3B in March 2002, the last servicing mission. Image courtesy STScI/NASA.*

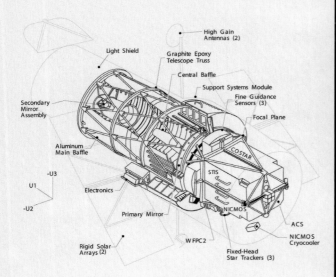

High Gain Antennas (2)

Light Shield

Graphite Epoxy Telescope Truss

Central Baffle

Support Systems Module

Fine Guidance Sensors (3)

Focal Plane

Secondary Mirror Assembly

Aluminum Main Baffle

COSTAR

-U3

STIS

U1

Electronics

NICMOS

-U2

Primary Mirror

WFPC2

ACS

NICMOS Cryocooler

Rigid Solar Arrays (2)

Fixed-Head Star Trackers (3)

January 2006, a supply feeding electrical power to the detectors failed, and then more recently, in January 2007, an electrical fuse blew, probably as a result of a short circuit within the ACS electronics.

And so this fifth, and final, servicing mission is critically needed to renew Hubble and ensure its longevity. The mission will be the heaviest servicing mission to date. Atlantis will be carrying about 22,000 pounds of hardware. Four carriers inside the shuttle cargo bay will carry the new science instruments, replacement hardware, tools for the astronauts, and the rig to attach Hubble to the shuttle. One of the carriers incorporates an advanced design and composite materials to save weight, allowing more material to be carried to orbit. A contingency shuttle will be on the pad and ready for an immediate launch and rendezvous with Atlantis in case this is needed to protect the crew.

The mission will be arduous for the astronauts, as they will be called upon to complete numerous space repairs, including ones that have not been attempted before. Currently, it's envisaged as

an 11-day flight. On the third day following launch, Atlantis will rendezvous with Hubble. Five separate space walks will then be needed to accomplish all of the mission objectives, with four astronauts undertaking extravehicular activities (EVAs) in pairs on alternating days. An example of a planning schedule for the five EVA days is shown on facing page. This plan is an early one, but it does include repairs to both ACS and STIS. As the timings of each activity are improved and as the priorities get established, the schedule will undoubtedly change, but it shows a mission timeline already stocked to the gills with important work.

There are three major goals for this complex return mission to Hubble. The first is to install a new camera and a new spectrograph: the Wide Field Camera 3 (WFC3) will replace WFPC2 and the Cosmic Origins Spectrograph (COS) will

■ **Above:** *A cutaway illustration of the Hubble Space Telescope showing the locations of the major subsystems and scientific instruments, shown as they are prior to SM4. During the fifth and final servicing mission, the Wide Field Camera 3 (WFC3) will replace WFPC2 and the Cosmic Origins Spectrograph (COS) will replace COSTAR. Complex repairs will be carried out to (hopefully) restore STIS and ACS to full operational capability. Image courtesy STScI/NASA.*

■ **Above right:** *On flight day 7 (EVA4) during SM3B in March 2002, astronauts Jim Newman (anchored to the end of the shuttle's 15-meter long robot arm) and Mike Massimino install the new Advanced Camera for Surveys (ACS) in place of the Faint Object Camera in the compartment at the base of the Hubble Space Telescope. Image courtesy STScI/NASA.*

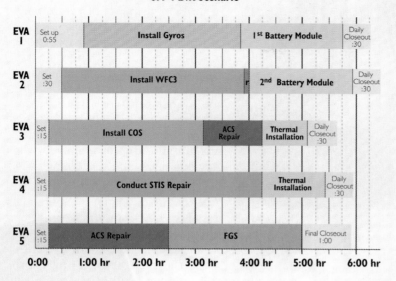

SM-4 EVA Scenario

| EVA 1 | Set up 0:55 | Install Gyros | | 1st Battery Module | | Daily Closeout :30 |

| EVA 2 | Set :30 | Install WFC3 | | r | 2nd Battery Module | | Daily Closeout :30 |

| EVA 3 | Set :15 | Install COS | | ACS Repair | Thermal Installation | Daily Closeout :30 |

| EVA 4 | Set :15 | Conduct STIS Repair | | | Thermal Installation | Daily Closeout :30 |

| EVA 5 | Set :15 | ACS Repair | | FGS | | Final Closeout 1:00 |

0:00   1:00 hr   2:00 hr   3:00 hr   4:00 hr   5:00 hr   6:00 hr

replace COSTAR, which is no longer needed to correct Hubble's vision. These installations will enhance Hubble's scientific capabilities.

The second goal is to restore the existing scientific capability through in-situ attempts to repair the failed STIS and ACS. This work will be very challenging. It would be hard enough on the ground with experienced technical staff, and so the planning for on-orbit repairs has to be very carefully planned and thought through. Although NICMOS was repaired on-orbit through the addition of a new cooling system, the STIS and ACS repairs will be much more difficult because they require the removal of numerous screws and anchors, and the replacement of electronic boards.

The third goal is to ensure the life extension of Hubble through increasing its hardware redundancy. Astronauts will replace all six of the telescope's gyroscopes, a fine guidance sensor, and all six batteries. New thermal coverings to Hubble's exterior will be added, and, in preparation for the de-orbit mission at the end of Hubble's life, a capture mechanism to the aft bulkhead will be attached.

There has been a history of innovation in Hubble instrument design, reaching back to the first generation of instruments, and the new instruments for SM4 are no exception. But it takes time for instruments to be selected, to be built, thoroughly tested, and then readied for launch. In the case of the COS, for example, it will be over 12 years between inception and launch, although the intervening cancellation exacerbated the process. The first opportunity to

build a new instrument for SM4 was announced in spring 1996, and by late 1997 the Cosmic Origins Spectrograph (COS) had been peer reviewed and selected.

The lead investigator, Dr Jim Green of the University of Colorado, has emphasized high efficiency as a primary design goal for this spectrograph. To achieve this, after the entrance aperture, the spectrograph employs a single reflection at the grating before incoming photons are captured by a state-of-the-art photon counting detector. The grating has an aspheric concave surface figure specified to compensate for Hubble's spherical aberration. Holographically generated grooves provide dispersion and correct the astigmatism, and ion-etching creates a blaze that optimizes efficiency over the chosen wavelengths. The detector is a windowless microchannel-plate array, with an opaque CsI photocathode and delay line readout developed by Dr Oswald Siegmund, University of California. Its faceplate is bowed to match the curvature of the incoming dispersed lightbeam.

The COS instrument has a design heritage that can be traced back to first and second generation Hubble spectrographs as well as to the successful FUSE spacecraft, which used a similar detector. COS makes use of returned space hardware, employing the original optical bench from the Goddard High Resolution Spectrograph (returned during SM2 in 1997).

The second instrument being prepared for SM4 is a new camera, called the Wide Field Camera 3 (WFC3). This instrument is being built as a facility class instrument to ensure that Hubble has a first-class imaging capability through the end of mission life. It is being developed at GSFC, and Drs Edward Cheng (Conceptual Analytics), Randy Kimble (NASA), and John MacKenty (Space Telescope Science Institute) have played important roles in getting the instrument designed, developed, and tested.

■ *Above: An early EVA timeline for SM4 is shown here as an illustration of the mission complexity. The timings and sequences of each activity will change as launch approaches. For example, this planning scenario schedules the installation of WFC3 ahead of COS, and it includes both the STIS and ACS repairs. Image courtesy STScI/NASA.*

Like COS, the WFC3 makes use of returned space hardware, this time from the original Wide Field and Planetary Camera.

Innovation is in evidence for WFC3, too. Ambitiously, this camera has two wide-field cameras, one that operates in the near ultraviolet and optical wavelengths, and the other in the near infrared, from 1 to 1.7 microns. Both channels are equipped with a wide assortment of filters, including low spectral-resolution prisms or grisms. While the ACS emphasized the red wavelength region, the WFC3 is considered a fully panchromatic camera, and provides excellent response with its enhanced CCD coatings from 200 to 900 nanometers. The infrared channel, utilizing the latest Mercury-Cadmium-Telluride (HgCdTe) detector technology should provide a vast sensitivity improvement over NICMOS. The infrared camera has a much larger field of view than NICMOS, which will be used to advantage in galaxy and quasar surveys.

Developing any new instrument for Hubble is a costly and painstaking process. Typically, a small group of scientists and engineers lead the development in conjunction with industry partners, with the overall project being funded and managed by NASA civil servants who can call on their own cadre of scientists and engineers. Staff at the Space Telescope Science Institute support the development program and take over routine science operations and calibration once the instrument is deployed. The Ball Aerospace and Technologies Corporation, in Boulder, Colorado, have played a major industrial role in developing many of the science instruments for Hubble. Their continued involvement throughout the program has facilitated the use of common software, systems, and manufacturing processes, from one generation of instruments to the next. Each generation of instruments has provided substantial improvements over previous ones, and these new SM4 instruments are no exceptions.

Without a doubt, the most complex EVA activities will be attempts to repair both STIS and ACS. Although both repairs await the final go-ahead from NASA management, a considerable engineering effort is in progress to demonstrate the feasibility of repairing these two important science instruments. A number of unique astronaut tools are being developed to help the astronauts remove covers and insert electronic boards. The tools and astronaut procedures are being rigorously tested using high-fidelity Hubble mock-ups on the ground to avoid surprises during the EVA itself, like a shelf or a tricky corner preventing access to a screw or a panel.

For the STIS repair, an astronaut will attempt to access the electronics box containing the failed component once the axial-bay doors in the aft shroud are opened. First, 111 fasteners that hold the electronics box cover in place must be removed. Engineers have already designed the special tool that will be attached to the existing cover plate and which will serve to capture the fasteners as they are removed. An astronaut will use a hand-held electric driver to unscrew each fastener. When the cover is removed, the astronaut will use other specially designed tools to remove the circuit board with the failed component and insert its replacement. A new cover plate will be installed and secured using two clamp-like latches (instead of 111 fasteners!).

The ACS repair will benefit from the planning that has already gone into STIS over the last couple of years. The repair will seek to restore and even improve the scientific performance of the wide field camera of ACS. The new electronics are based on technology in development for the James Webb Space Telescope. During the proposed repair the ACS will remain in place. The access cover will be removed using a STIS-style capture plate, and a grid has to be removed with a shearing tool before the failed electronic

■ **Above:** *During SM4, the Cosmic Origins Spectrograph (COS) will be installed in place of COSTAR, which will no longer be needed to correct Hubble's vision. High efficiency was a primary design goal for this new spectrograph. This external view of the COS instrument shows the connector interface panel. Image courtesy STScI/NASA.*

fourth gyro failed, leaving only two working. While the image quality has not been degraded, observations with either orientation or timing constraints are now more difficult to schedule, and only about half the sky is observable at any one time (compared to greater than 80 percent of the sky in three-gyro mode.) The observatory will return to three-gyro operations after SM4.

Astronauts will replace all six of Hubble's original nickel-hydrogen batteries, which have provided electrical power during approximately one-third of the time Hubble spends in the Earth's shadow when the solar arrays do not function. Sixteen years into the mission, these 125-pound batteries have lasted more than 11 years longer than their design lifetime, longer than the batteries in any other spacecraft in low Earth orbit. In addition to the stability of nickel-hydrogen chemistry, this durability attests to the robust design of these batteries and their careful on-orbit management by NASA engineers.

The new batteries are made using a process called wet slurry, which makes them physically stronger and better performing. Each new battery has an added safety feature: a battery isolation switch that ensures no electrical power is present at the connectors while the switch is in the off position.

Astronauts will add stainless steel sheets at various locations on Hubble's exterior to help control the telescope's internal temperature.

boards can be replaced. A new power supply will be clamped to a handrail to complete the installation. The total EVA time has to be less than four hours to fit into the timeline. Engineers and scientists are busy working on this new repair task, and only time will tell whether it is truly feasible during SM4.

The gyros that astronauts intend to replace on SM4 are used to control the pointing of the telescope during science observations. The original telescope design called for three of the six gyros to be operating at a time. To conserve the use of the three gyros currently remaining, a new attitude-control system was devised requiring only two operating gyros. This enabled one to be turned off in August 2005, and Hubble observed successfully in two-gyro mode, until September 2007, when a

■ **Above:** *During preparations for SM4, a new camera called the Wide Field Camera 3 (WFC3) – seen towards the top of this view – is lowered vertically into its protective enclosure for acoustic testing. This instrument is being built as a facility class instrument to ensure that Hubble has a first-class imaging capability through the end of mission life. The WFC3 is being developed at NASA's Goddard Space Flight Center. Image courtesy NASA Goddard Space Flight Center and STScI.*

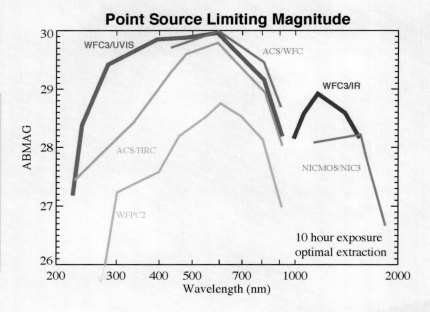

Point Source Limiting Magnitude

(Axis labels: ABMAG (vertical), Wavelength (nm) (horizontal); curves labeled WFC3/UVIS, ACS/WFC, WFC3/IR, ACS/HRC, NICMOS/NIC3, WFPC2; annotation: 10 hour exposure optimal extraction)

■ **Right:** *A comparison of the point-source sensitivity of the four main Hubble cameras: WFPC2, NICMOS, ACS and WFC3. The plot shows the magnitude of the faintest star that could be detected in a total exposure time of ten hours, using the most sensitive filters available over the full wavelength region of the instruments. Larger magnitudes correspond to fainter sources. The WFC3 will be able to detect a star similar to our Sun in a ten hour exposure at a distance of several million light-years. Image courtesy T. Brown, STScI.*

These sheets will cover portions of the existing multi-layer insulation that have degraded over time due to exposure to the harsh environment of space.

When Hubble reaches the end of its life, NASA plans to de-orbit it safely into the open ocean by firing a rocket that will be attached on a future robotic visit to the observatory. To prepare for this mission, engineers developed the Soft Capture Mechanism (SCM) and the Rendezvous Navigator Sensor (RNS). The SCM is a ring-like device that will be attached to Hubble's aft bulkhead. It will provide a mating interface and navigational targets for future rendezvous, capture, and docking operations. The RNS system consists of optical and navigation sensors, as well as supporting avionics and processors. However, it will not be installed on Hubble – its role during SM4 is to collect data during the capture and deployment of the telescope. This information will be used for developing the navigation systems of the future spacecraft that will de-orbit Hubble at the end of its useful life.

The improvements of SM4 are expected to keep Hubble operational until at least 2013. They will also greatly enhance the discovery power of the observatory. For example, both COS and WFC3 contain advanced technology that far surpasses what has been available on Hubble to date, and key performance factors are expected to improve by 10 to 70 times.

Astrophysicists are eagerly awaiting the arrival of COS. Spectroscopy is such an important tool for quantitative measurement that its absence from Hubble for the past few years has been a severe blow. Our understanding of the origin and evolution of the cosmos depends critically upon our ability to measure the total mass, the distribution, motions, temperatures, and composition of atoms and molecules in the universe. COS is a powerful instrument for a broad range of science problems because its high efficiency will allow observations of very faint targets, like high-redshift quasars, and yet its versatility will also allow us to probe the nature of nearby galactic objects, like extra-solar planets. A primary task of the COS will be to measure the ionization and baryon content of the intergalactic medium and the origin of large-scale structure in the Universe, through observations of quasars shining through material in the 'cosmic web'.

The WFC3 will continue to allow Hubble to obtain the spectacular images that have become such a hallmark of the telescope. The power of this camera lies in its ability to record light over such a broad wavelength range from the ultraviolet through to the near infrared and in the superb low-noise characteristics of its detectors. Over time this new camera will help to augment the results obtained from previous generations of cameras, whilst its new infrared eyes will prepare the groundwork for the next big space venture, the 6-meter James Webb Space Telescope. If STIS and ACS can be repaired to complement these new instruments, then Hubble will be at the peak of its power. Be ready to sit back and enjoy the discoveries!

"For astronauts, and for me on a personal level," writes John Grunsfeld, "the opportunity to leave planet Earth and fly in the space shuttle on any mission is an incredible privilege. It is nevertheless important to remember that this activity carries with it more than a significant risk, as the tragic loss of the Columbia and the

■ *Hubble's view of a nearby barred spiral galaxy, NGC 1672. The image unveils details in the galaxy's star-forming clouds and dark bands of interstellar dust. This galaxy is more than 60 million light-years away. This image was constructed from data taken with the ACS. Image courtesy NASA, ESA, and The Hubble Heritage Team (STScI/ AURA)-ESA/Hubble Collaboration.*

STS-107 crew patently demonstrated. As a crew, we go forward with the Hubble Servicing Mission 4 with the knowledge that we are participating in an activity that is much bigger than all of us and worth the risk to ourselves and our families. The work that we do on-orbit has become an integral part of our culture. When we go to Hubble in 2008, we will not just write another chapter in the Hubble story, but will extend a work that is already a major mark in history."

For further information, consult these websites:

Main Hubble Space Telescope website:

http://hubblesite.org/

The Telescope and Where's Hubble Now?

http://hubblesite.org/the_telescope/where.a.s_hubble_now/basic_version.php

Cosmic Origins Spectrograph (COS) site:

http://cos.colorado.edu/

Far Ultraviolet Spectroscopic Explorer site:

http://fuse.pha.jhu.edu/

NASA's Wide Field Camera 3 (WFC3) site:

http://wfc3.gsfc.nasa.gov/index.php

J. Chris Blades is a tenured astronomer with AURA and is currently Project Scientist for SM4 at the Space Telescope Science Institute. Following his Ph.D. from University College, London, he worked at the Anglo-Australian Telescope and for the International Ultraviolet Explorer, and has spent the past 25 years at STScI.

11

The History and Future
OF TELESCOPES...

Jerry Nelson of the University of California Observatories
(University of California at Santa Cruz), the pioneer of the segmented
mirror design that led to the construction of the 10-m Keck telescopes,
shares his views of the past, present and future of telescopes to mark the
four hundred year anniversary of its invention.

Image courtesy TMT Observatory Corporation.

...400 Years from Galileo to
KECK
and
BEYOND

HUMANS ARE inquisitive and visual creatures. What we 'see' is probably our most important sensory input, and much of what we learn and know is expressed in visual images and expressions. Our objective understanding of our environment comes from pictures of it, including what we can see with the naked eye, what we can see with microscopes and what we can see with telescopes.

Modern and future telescopes do much more than take 'pictures'. In practice modern telescopes collect information about our universe that is often pretty abstract. We detect photons, counts, wavelengths, and then process this information to form visual expressions of the information so we can more readily absorb it.

Telescopes allow us to:

- **see objects that would otherwise be too faint to detect with our naked eye**
- **see objects at much higher angular resolution than we could otherwise detect with our naked eye**
- **see objects in wavelengths other than those to which our eyes are sensitive.**

History of the Telescope

Spectacle lenses have been in use for hundreds of years, but their first application to making a telescope is usually credited to a Dutch spectacle maker, Hans Lippershey (c1570-c1619), probably around 1608. It is not clear who actually first discovered this application of lenses, and it may have been found by several others. Upon hearing about this device, Galileo, the Italian instrument maker, reinvented it, and in 1610 published his night sky observations. These

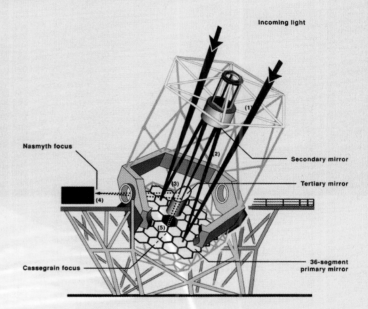

included the discovery that the Moon was not smooth but was covered with craters, that Jupiter had four moons which circled it, and that the Milky way was composed of millions of stars. The discovery that Jupiter had moons that circled it was particularly revolutionary, since the accepted dogma was that all objects circled the Earth (geocentric view). He named his device a telescope, from the Greek words for *seeing at a distance.*

■ *Above: The author of this article, Jerry Nelson, was the pioneer of the segmented mirror design that led to the construction of the 10-meter Keck telescopes. This diagram shows the path of the incoming starlight (1), first on its way to the Keck's segmented primary mirror; reflected off the primary, toward the secondary mirror (2); bouncing off the secondary, back down toward the tertiary mirror (3); and finally reflected either off the tertiary mirror to an instrument at the Nasmyth focus (4) at the side, or to the Cassegrain focus (5) beneath the primary mirror. Diagram courtesy W.M. Keck Observatory and California Association for Research in Astronomy.*

Infrared

Visible

This telescope consisted of two lenses, an objective that gathered and converged the light, and a diverging eyepiece lens that produced parallel light that could enter the eye and be focused onto the eye's retina. Several types of refracting telescopes can be built, but all use the principle of refracting (bending) the light that enters a lens to allow images to be made. All of the early astronomers used refracting telescopes, including Galileo, Kepler, and Huygens.

These early telescopes brought two great advantages: they magnified the image so greater angular resolution was achieved than was possible with the naked eye; and they gathered more light than the naked eye could collect, thus allowing observations of fainter objects. The naked eye

■ **Above:** *The Eta Carina star-forming region is seen in infrared light in this image by the Spitzer Space Telescope. Infrared light is detectable from deep within the dark clouds whereas visible light is emitted from only the surface of the clouds. The inset image shows the same region in visible light. Inside the cloud, dense cores of dust and gas form finger-like tendrils whose outer surfaces are illuminated by ultraviolet light from recently formed hot stars. Main image courtesy NASA/JPL-Caltech/N. Smith (Univ. of Colorado at Boulder). Inset visible light image courtesy NOAO.*

pupil diameter is small: about 1mm in bright daylight and up to 7mm at night. Lenses can be made much larger than 7mm and this increase in size allows the collection of more light from a faint object.

There is a fundamental limit to angular resolution caused by diffraction. Because light acts as a wave, there are interference effects, and the best angular resolution an optical system can deliver is about λ/D, where λ is the wavelength of light and D is the diameter of the optical system. For a 1mm pupil (our eye) this resolution limit is about 1 arcminute or 1/30th of the diameter of the Sun or Moon. So, the value in magnifying the images for Galileo was dependent on the fact that the diameter of his telescope was much greater than the size of his eye's pupil.

In the 1670s, Isaac Newton explained the chromatic aberration found in these refracting telescopes and built a small reflecting telescope (using mirrors rather than lenses) to demonstrate the principle (and also the avoidance of chromatic aberration). Various technical difficulties slowed its use, and only in the 19th century did reflecting telescopes come into common usage and begin to dominate the astronomical telescope scene.

11 - The History and Future of Telescopes . . . 400 Years from Galileo to Keck and Beyond

Telescopes at Different Wavelengths

Telescopes through the 19th century were used to study light that was detectable by the human eye or by photographic film. Visible light covers the wavelength range of $0.4\mu m$ to $0.7\mu m$ (μm = micrometer or micron). In the 20th century it was recognized that a good deal of infrared radiation also was given off by celestial objects and could be detected through the Earth's atmosphere. Infrared typically ranges between 1-$30\mu m$. It was only in the 1960s that IR detectors matured enough to make infrared astronomy an interesting astronomical field. In the 1930s, celestial radio sources were discovered and radio astronomy was born. Radio astronomy typically covers wavelength regions from 1mm to 1 meter.

Infrared Telescopes

Many optical telescopes can also serve as infrared telescopes, but many parts of the electromagnetic spectrum are absorbed by the atmosphere, so the most powerful infrared telescopes have been planned for space. Pathfinding astronomy in the infrared was led by the Infra-Red Astronomical Satellite (IRAS) launched in 1983. A 0.6-meter reflecting telescope cooled to 4 K, it carried out a survey of the entire celestial sky, and made many fundamental discoveries, including starburst galaxies. It operated over a wavelength range 8-$120\mu m$, a region largely absorbed by our atmosphere. More recently (2003) the Spitzer Space Telescope was launched, and it covers a wavelength range of 3-$180\mu m$, is cooled to 4 K and has a 0.9-meter primary mirror. Its photon detectors are much more capable than those of IRAS and it has made wonderful new discoveries. An example of Spitzer's imaging capability is well illustrated by its view of the Eta Carinae nebula, which uses representative color to portray radiation from 3, 5, and $8\mu m$.

Because of the expansion of the universe, the most distant objects (seen when they were very young) are dominantly radiating in the infrared (their light has been greatly redshifted). For these studies, NASA is embarking on the construction of the James Webb Space Telescope (JWST) due to be launched in 2013. This 6-meter diameter telescope will be cooled to ~ 30 K and give superlative infrared performance in the 2-$30\mu m$ wavelength range. It is notable that because of its great size, the primary mirror is composed of 18 smaller hexagonal segments, allowing the mirror to be folded for launch.

A computer-generated model of the James Webb Space Telescope illustrates the 18 segments making up the 6.5-meter mirror and the large sails that act as an insulator, maintaining the telescope and science instruments at very cold temperatures. All the science instruments are located behind the main mirror. Image courtesy NASA/NGC.

Radio Astronomy

Radio astronomy has blossomed into a very important field and in the 20th century many large radio telescopes were built. The largest steerable dish in the world is the Green Bank Telescope, with a 100-meter diameter dish antenna, completed in 2000.

In radio astronomy the wavelengths are so long that it is possible and advantageous to combine arrays of smaller telescopes into interferometers. The most successful of these to date is the Very Large Array in Socorro, New Mexico. The VLA is the cornerstone of the Very Long Baseline Array (VLBA), a network of radio telescopes that spans the US continent. Building on these successes, the US and Europe are currently building the Atacama Large Millimeter Array (ALMA) interferometer in northern Chile at an altitude of 5000 meters. It will be operational in 2012 and consists of fifty 12-meter telescopes and twelve 7-meter telescopes. Working at wavelengths bracketing long wavelength infrared and short radio waves, ALMA will open the sub-millimeter universe.

Above: The Green Bank Radio Telescope, the world's largest steerable dish, is located in West Virginia. The GBT is part of the National Radio Astronomy Observatories (NRAO) which also operates the VLA at Socorro, New Mexico. Image courtesy NRAO/AUI.

Back to the Future

We now return to describe modern and future ground-based optical-infrared telescopes. In the 1930s the Palomar 5-meter telescope was begun, and finished in 1948. It was the world's largest until 1993, when the 10-meter Keck Telescope went into operation. In the 1970s and 80s, various technologies developed that made radical innovations over the Palomar telescope possible.

One key idea was the development of segmented mirrors. Traditionally, the primary mirror of an optical telescope was an enormous challenge because:

- *The mirror blank was extremely difficult to make*
- *The mirror was flexible, requiring sophisticated supports to retain its shape*
- *Polishing was difficult and very time consuming due to the large size*
- *Reflective coatings required large vacuum chambers to enclose the mirror*
- *Handling of the nearly priceless mirror was difficult, awkward, and expensive*
- *The mirror was thick and massive, making the telescope massive as well.*

With a primary mirror composed of segments, the key challenges changed dramatically. Segments are unusually difficult to polish since they are off-axis sections of the parent parabolic mirror. The segments must be very carefully aligned relative to each other in order to behave as though it is a single mirror. The array of segments must be actively stabilized over time (as temperature changes) and as the telescope points around the heavens, as the steel support system will deform by many times the allowed misalignments (typically tens of nanometers) due to changing gravity forces.

These challenges have been met in various ways, and the result is two 10-meter Keck telescopes, a third 10-meter telescope being built in the Canary Islands (the Gran Telescopio Canarias (GTC)), and two 8-meter telescopes built with spherical primaries (the Hobby-Eberly Telescope (HET) in Texas and the Southern African Large Telescope (SALT)).

The Keck telescopes each have 36 hexagonal segments that make up their primary mirrors. The two telescopes are identical, and Keck 1 began regular science operations in 1993, while Keck 2 began science operations in 1996. They were built and are operated by the University of California and Caltech.

Other astronomers also pressed the technology frontiers to build 8-meter telescopes, whose success relied on making and controlling the shape of very thin 8-meter diameter mirrors. The control problem, as with the Keck telescopes, relies heavily on modern computer and control technologies unavailable 30 years ago. There are now seven 8-meter telescopes. The Europeans built four 8-meter telescopes in Chile, creating the Very Large Telescope. The Japanese built an 8-meter telescope (Subaru), and a consortium involving the USA built two Gemini 8-meter telescopes, one located in Hawaii, and another in Chile. Yet another approach, using thick but lightweight monolithic mirrors is being used in the almost complete Large Binocular Telescope (see article 'Unveiling A New Generation of Extremely Large Telescopes ... The Large Binocular Telescope Achieves First Light' by Richard Green and John Hill in last year's volume, *State of the Universe 2007*).

■ *Above: A computer-generated scene shows the ALMA array deployed in the high Chilean desert, well above water vapor that blocks submillimeter wavelengths of light. When complete, the ALMA array will be the world's largest imaging array in the world, using 64 steerable dishes, each 12 meters across. ALMA will give astronomers the opportunity to peer deep into obscured regions of space, such as star-forming regions, and the core of our Milky Way thought to harbor a supermassive black hole, at an unprecedented level of detail. Image courtesy NRAO/AUI and computer graphics by ESO.*

■ *Above right: The twin 10-meter Keck telescopes are shown in this computer generated cutaway rendition. The W. M. Keck Observatory located on Mauna Kea, Hawaii, houses the largest telescopes in the world. Image courtesy Tom Connell, Wildlife Art\Weldon-Owen, Inc.*

telescope has nine times the collecting area of Keck, it has many more segments. Keck has 36 1.8-meter diameter hexagonal segments, while TMT has 492 1.4-meter segments.

Making and properly aligning the segments are the biggest challenges in constructing a giant telescope. The ensemble of segments should behave as a single continuous paraboloid. Thus the individual segments are not axi-symmetric about their own centers.

These off-axis segments represent a challenging polishing problem. Polishing is effective when the polishing tool fits the surface well. Thus spherical surfaces are fairly easy to polish. These non-spherical segments can be polished either with small tools (small enough that the tool approximately fits the optical surface where it is located) or, as was done with Keck, by using a large spherical tool and deforming the mirror blank with carefully applied forces, so it is deformed into a spherical shape.

The Future

Astronomers need to gather light from the very faint targets of interest in order to understand them and their environments. As stated earlier, the key capabilities needed are great light gathering power (mirror collecting area) and angular resolution (set by the diameter of the telescope primary mirror).

Around 2000, several groups of astronomers began to seriously explore the science potential, the technological challenges, and the costs of making truly giant telescopes, telescopes greater than 20 meters in diameter.

There are two technologies being explored for making such giant telescopes: segmented mirrors along the lines of Keck Observatory, and a small group of 8-meter diameter mirrors polished and mounted to form a single telescope.

In 2000, the University of California and Caltech began the study of a 30-meter telescope named the California Extremely Large Telescope (CELT). This project has evolved and is now the Thirty Meter Telescope (TMT), and has Canada as an additional partner.

TMT has learned many lessons from Keck and is designed using many of the successful ideas seen in the Keck Observatory. Because the

■ **Above:** *The Thirty Meter Telescope is shown in this computer-generated illustration. A centrally-located tertiary mirror directs light to the Nasmyth focus, located along the horizontal axis of the mount. Image courtesy TMT.*

■ **Top left:** *The Sun sets over this aerial view of the Paranal peak in Chile, housing the four giant 8.2-meter telescopes of the Very Large Telescope. YEPUN is in the front, and from left to right in the background are ANTU, KUEYEN and MELIPAL. Some of the concrete supports for the smaller, movable telescopes of the VLT Interferometric Array and the rails on which they will move, are seen to the left. The central, mostly subterranean laboratory in which the light beams from all the telescopes will come together at the interferometric instruments is about halfway between ANTU and YEPUN. Image courtesy ESO.*

■ **Middle:** *The Thirty Meter Telescope (TMT) will contain 492 separate hexagonal mirror segments, each controlled by individual actuators. The design expands on the successful operation of the 10-meter Keck telescopes. This diagram by the author shows the layout of the individual mirror segments. Image courtesy J. Nelson.*

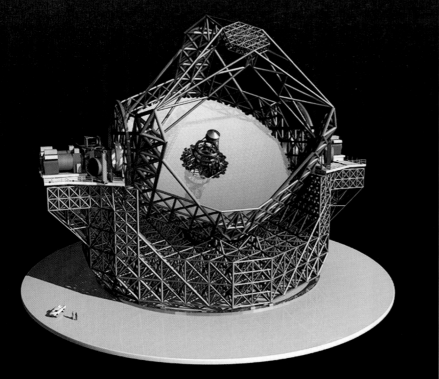

In the TMT, each segment edge will be equipped with a pair of capacitive sensors that can detect changes in the relative height of a segment against its neighboring segments. By measuring all these height differences (about 10 times per second), the overall relative positions of all the segments can be calculated, and using three displacement actuators per segment (that can make moves as small as 5 nanometers), the segments can be held in proper alignment.

The TMT telescope is built around this segmented primary mirror and directs starlight to either of two Nasmyth platforms where the science instruments are located.

In this case the large tool and the spherical mirror blank fit well and the entire segment can be polished at once.

Aligning the mirror segments to their proper orientations and keeping them in place is another challenge. Alignment is tested by using starlight and carefully designed test equipment that can detect very small tip-tilt errors and also detect relative piston errors that cause discontinuities from segment to segment. Such discontinuities can be measured to an accuracy of about 10 nanometers.

An even more ambitious segmented-mirror telescope is being planned by the European Southern Observatory (ESO). ESO is planning to build a 42-meter diameter telescope, the European Extremely Large Telescope (EELT), with segments roughly the same size as TMT. The optical design is interestingly different, however. TMT has a primary mirror, a secondary mirror and a third mirror (a fold flat) that sends the science light to the Nasmyth platforms. The ESO design has a five mirror system that allows a relatively small deformable

■ *Above:* European astronomy has received a tremendous boost with the decision from ESO's governing body to proceed with detailed studies for the European Extremely Large Telescope (E-ELT). This study, with a budget of 57 million Euros, will make it possible to start, in three years' time, the construction of an optical/infrared telescope with a diameter around 40 meters that will revolutionize ground-based astronomy. Image courtesy ESO.

■ *Right:* The Giant Magellan Telescope, scheduled for completion in 2016, will have the resolving power of a 24.8-meter telescope. Its seven 8-meter mirrors are shown in a specially-designed building. Large openings in the structure allow the outside air to flow evenly around the telescope, equalizing the temperature and reducing the effects of turbulence around the telescope. Image courtesy Giant Magellan Telescope – Carnegie Observatories.

mirror to be part of the telescope optics. This mirror is a key component of the telescope adaptive optics system. We shall discuss adaptive optics in a moment.

Another giant telescope with its own set of engineering challenges is the Giant Magellan Telescope (GMT), designed around the concept that large monolithic 8-meter mirrors can be used as the basic building block for a giant telescope. The GMT has seven circular 8-meter mirrors forming its primary mirror, creating a telescope with an aperture of 24.5-meters. This project is being led by the Carnegie Observatories of Washington in Pasadena, California.

These giant telescopes have many technical challenges in achieving their collecting area. However, probably their biggest challenge and greatest science impact will come from the use of adaptive optics.

Adaptive Optics Now and in the Future

Superior image quality is an essential aspect of the science potential of a telescope. Ground-based telescopes must look through the Earth's atmosphere, and the atmosphere blurs the images of stars. This effect is sufficiently serious that even large telescopes make images of stars no sharper than a 100mm diameter telescope!

The nature of the disturbance of the atmosphere is this: the starlight coming to the Earth can be understood as plane waves coming to us. This plane wave would be focused by a telescope to make a diffraction-limited image. However the atmosphere disturbs the plane wave, making some parts speed up a bit and some parts slow down a bit, thus crumpling the plane wave. This is caused by spatial variations in the turbulent atmosphere - small temperature variations cause small density variations, causing small variations in the speed of light, which in turn cause the plane wave to become crumpled and this produces the degraded images.

Adaptive Optics (AO) is a technique to compensate for this by introducing an optical aberration that is equal and opposite to the wavefront disturbance of the atmosphere, thus restoring the plane wave and the potential for diffraction-limited imaging.

There are several challenges in the application of the ideas of adaptive optics (AO). The first is measuring the disturbance caused by the atmosphere, the second is reconstructing the atmospheric wavefront error, and the third is correcting this wavefront error. All of these are made challenging by the fact that the atmosphere changes rapidly, requiring the above adjustments to be done hundreds of times per second.

The basic ideas of an AO system involves light from the telescope being directed to a deformable mirror and a wavefront sensor, which constantly senses the effectiveness of the corrections and

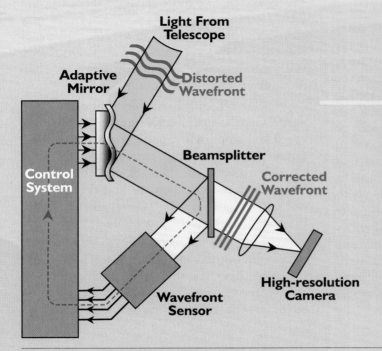

Adaptive Optics

The light from the telescope (typically science light and the light that is used to measure the atmosphere) reflects from a deformable mirror (that applies the desired correction to cancel the effect of the atmosphere), then some of the light is separated to enter the wavefront sensor and part proceeds onwards to the science instrument. The wavefront sensor effectively measures the residual wavefront error and sends this error to the deformable mirror. As shown this is a 'closed-loop' system where the sensor is seeing the residual wavefront error, and the control system is trying to drive this error to zero. Image courtesy J. Nelson, Center for Adaptive Optics.

Hubble Deep Field

■ *Utilizing adaptive optics to correct images for the blurring of the atmosphere, a 30-meter telescope will be able to produce images 12.5 times sharper than they are achieved with the Hubble Space Telescope. These sample images show the amount of additional detail that such an improvement in resolution would reveal in a distant galaxy. Images courtesy J. Nelson, TMT and CELT.*

30mm + adaptive optics resolution

sends the result back to the deformable mirror in a closed loop, as shown in the panel.

AO systems such as this are currently in use on several 8-10 meter telescopes today, and can deliver diffraction-limited images in the near infrared (wavelengths 1-2.5μm).

A challenge for all astronomical AO systems is measuring the ever-changing atmosphere. Ideally one would like to use light from the science target, as this follows exactly the proper path through the atmosphere, but most interesting targets are too faint for this. Sometimes a nearby bright star can be used, but they must be quite close to the science target and they are thus scarce. A more powerful approach is to generate an artificial beacon that can probe the atmosphere. We are doing this at the University of California Observatories with lasers that excite sodium atoms in the mesosphere (~ 90km up in the atmosphere) which then re-radiate, producing a bright spot that emulates a star. The astronomer directs the laser to make an artificial beacon right in front of the science target, thus allowing measurement of the proper atmospheric column.

Typically astronomers measure the slopes of the wavefront errors across the aperture with Shack-Hartmann wavefront sensors. The slopes are then combined to provide an estimate of the wavefront error. The more densely the slopes are measured, the more accurate the wavefront estimate is, and the better the AO system may correct the atmosphere. On today's 8-10 meter telescopes slopes are measured at ~ 300 places. With tomorrow's 30 meter telescopes we shall measure at 3000-10,000 places.

Making deformable mirrors that can change shape with the spatial resolution needed, and the speed and amplitude needed, is a real challenge. Often astronomers use piezoelectric actuators to push and pull on the back of a thin mirror. These devices are expensive and delicate, and one of the challenges for AO on giant telescopes is building these adaptive mirrors.

Why are we doing all this? AO delivers diffraction-limited images, and for a 10-meter telescope this gives an image size improvement of a factor of 100! (For example, light from a single star is focused into an area 100 times smaller than previously possible, allowing finer details to be rendered visible). On a 30-meter telescope the improvement is 300! These huge improvements in image size bring us vastly increased knowledge about what we are studying, and equally important, allow us to study much fainter objects. Typically telescopes on Earth observing in the infrared must contend with large background signals from infrared light from the Earth's atmosphere and even from the warm optics of the telescope. It's analogous to doing optical astronomy during the day. So with sharper images, the amount of background signal that hides the star drops dramatically. A factor of 100 reduction in image size speeds up an observation by a factor of 10,000, because shorter exposure times are required for the more concentrated light.

These are heady times for ground-based astronomy. We expect in the next 10 years to see a 30-meter telescope built, and possibly more than one. This huge increase in collecting area will have a revolutionary impact on our study and understanding of the distant and early universe. Combined with the multiplicative power of adaptive optics, these giant telescopes will have completely unprecedented sensitivity and angular resolution, making the direct observation of extrasolar planets practical, measuring the first generation of galaxies in the universe, and studying the details of star and planet formation and much more.

Ideally we would like to build our telescopes in space where we can avoid the absorption and blurring caused by the atmosphere. However space is expensive to build in and difficult to get to. People are working hard to develop giant telescopes for space that will use some of the ground-based telescope technologies. Very thin segmented-mirrors with adaptive-optics-like controls will someday allow huge telescopes in space, made of very thin (low-mass) mirror segments whose shapes are actively controlled. Then we shall build 100-meter diameter telescopes (and more) in space.

For further information, consult these websites:

W.M. Keck Observatory:
http://www.keckobservatory.org/

Spitzer Space Telescope:
http://www.spitzer.caltech.edu/
spitzer/index.shtml

The James Webb Space Telescope (JWST):
http://www.jwst.nasa.gov/

The Thirty Meter Telescope (TMT)
http://www.tmt.org/

ESO home page:
http://www.eso.org/public/

Gemini Observatory:
http://www.gemini.edu/

The Giant Magellan Telescope (GMT):
http://www.gmto.org/

ESO OverWhelmingly Large Telescope (OWL):
http://www.eso.org/projects/owl/

Professor Jerry Nelson is an astronomer at the University of California Observatories and is Research Scientist for The Thirty Meter Telescope (TMT). He is the Keck Observatory Scientist most closely involved with the design of the two Keck telescopes, and is also involved in the adaptive optics system being used and developed at Keck.

12

Chasing a Runaway
UNIVERSE....

That the Universe is expanding is a well-known concept in astronomy. The question of whether it would eventually start contracting, or would continue expanding forever (though gradually slowing down), puzzled scientists. Then, in 1998, cosmology was turned upon its head, when it was discovered that the Universe's expansion is actually accelerating. One of the key players in this discovery, *Alexei V. Filippenko*, explains in plain English what a dramatic discovery this was and where it might all lead.

...Dark Energy and the ACCELERATING UNIVERSE

COSMOLOGY, THE subset of astronomy that deals with the structure and evolution of the Universe as a whole, is one of the grandest of all subjects. Cosmologists are interested in the age of the Universe (when did it begin?), in its ultimate fate (will it expand forever or collapse?), and in its overall extent and shape (is it infinite, or does it wrap around like a balloon?). They also want to understand the formation and evolution of galaxies like our Milky Way; these are the fundamental building blocks of the Universe. Ultimately, we wish to understand our origin, and our connection with the cosmos as a whole. These are topics about which humans have probably wondered ever since developing sufficiently high intelligence and curiosity.

■ **Above:** The Hubble Ultra Deep Field, imaged with the Hubble Space Telescope. This is a tiny fraction of the sky, roughly the apparent size of a small pebble held at arm's length, yet it contains several thousand galaxies. Based on images such as this, astronomers estimate that there are about 100 billion galaxies within the observable part of the Universe – and the observable part (about 14 billion light-years in all directions) may be just a small fraction of the entire Universe. Image courtesy NASA, ESA, S. Beckwith and the Hubble Ultra Deep Field Team (STScI).

■ **Left:** Hubble Space Telescope image of the pinwheel-shaped galaxy NGC 1309. Our Milky Way Galaxy probably looks similar to this typical spiral galaxy, when seen from the outside. Our Galaxy is about 100,000 light-years in diameter and contains about 200 billion stars, gravitationally bound together. New stars have recently formed (and are still forming) in the spiral arms. Image courtesy NASA, ESA, The Hubble Heritage Team (STScI/AURA), and Adam Riess.

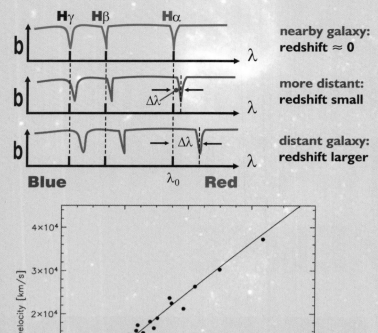

nearby galaxy:
redshift ≈ 0

more distant:
redshift small

distant galaxy:
redshift larger

Blue λ_0 **Red**

The Expansion of the Universe

A central figure in cosmology was Edwin Hubble, after whom the famous Hubble Space Telescope is named. He made a number of important discoveries, but most central to this article is his evidence that all but the nearest galaxies (which are gravitationally bound to the Milky Way Galaxy) are moving away from us – the Universe is expanding. Hubble made this discovery by passing the light from galaxies through a prism, producing a rainbow, or 'spectrum.' The spectra of galaxies were shifted toward redder (longer) wavelengths, and this 'redshift' was observed to be larger for the more distant galaxies than for the nearby galaxies.

Let's examine this result more closely. If we plot the brightness of dispersed light (such as that of a rainbow) as a function of color, or wavelength, we get a graphical representation of a spectrum, and from analysis of this we can determine interesting quantities such as the chemical compositions and temperatures of the typical stars in a galaxy. But we can also determine the motion of the galaxy toward or away from us by measuring whether its light is shifted to bluer or redder wavelengths (respectively), similar to the audible Doppler effect. The amount of blueshift or redshift is proportional to the galaxy's speed, v.

In the expanding Universe, the redshifts are actually produced by the expansion of space itself, not a motion of galaxies (like bullets) through a pre-existing space, but this is a subtle difference, and the formula relating redshift and recession speed turns out to be the same as that of the Doppler effect. Converting redshifts to speeds for many galaxies at various distances, a modern-day version of Hubble's diagram shows that his original relationship works to the farthest reaches of the observable universe.

A pictorial representation of Hubble's results shows that galaxies are moving away from our Milky Way Galaxy, and at the present time, their speed of recession is proportional to their distance from us.

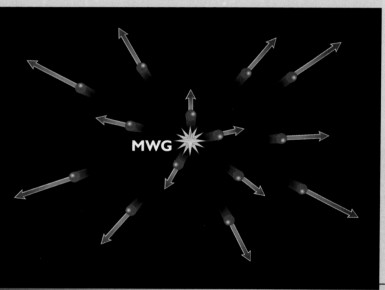

■ **Top:** *Idealized spectra of three galaxies at different distances from us. Brightness (b) of the light is plotted against wavelength (λ), or color. The main optical (visible) absorption lines of hydrogen are shown. In the nearby galaxy, these lines fall at roughly the same wavelengths as measured in a laboratory gas at rest. The more distant galaxy exhibits the same lines, but shifted to longer (redder) wavelengths. In the most distant galaxy shown, the redshift is even greater. If one measures the shift in wavelength, $\Delta\lambda = \lambda - \lambda_0$ (where λ is the measured wavelength of a line and λ_0 is its rest wavelength), then the redshift (z) is defined to be $\Delta\lambda/\lambda_0$, and the speed of recession (v) can be computed from $v \approx cz = c\,\Delta\lambda/\lambda_0$, the usual Doppler formula, where c is the speed of light. Image courtesy Alex Filippenko and Frank Serduke.*

■ **Middle:** *The Hubble diagram for several dozen galaxies, based on modern data. Distances are given in units of megaparsecs (Mpc), where 1 Mpc is equal to about 3.26 million light-years. Edwin Hubble's original data fell into the tiny square in the lower-left corner of this plot. The linear relationship*

*between galaxy speed of recession (v) and distance (d) is known as Hubble's law, $v = H_0 d$, where H_0 is the present-day value of Hubble's constant. (The value of H changes with time, yet is constant throughout the Universe at any given time.) Note that Hubble's law by itself does **not** imply that the speed of a given galaxy increases with time; rather Hubble's law states that at any given time, distant galaxies move away from us faster than do nearby galaxies. Image courtesy Saurabh Jha and Alex Filippenko.*

■ **Bottom:** *The uniform expansion of the Universe in two dimensions, as seen from the perspective of our Milky Way Galaxy (MWG); the third dimension is not shown, for clarity. Galaxies are moving away from the MWG, and at the present time, their speed of recession is proportional to their distance. Image courtesy Alex Filippenko and Frank Serduke.*

But before we move on, we should note that there is something very strange about this diagram: galaxies seem to be receding only from *us*, the Milky Way! Do they not like us, or is it something we said? Are they lactose intolerant? Are we in a central, preferred position – the center of the Universe?

Such a view, of course, would be quite anti-Copernican and hence unlikely, but one can also argue against it by noting that in any uniformly expanding space, all galaxies move away from all others. A hypothetical one-dimensional universe consisting of an expanding rubber tube, with ping-pong balls (galaxies) on it, has this property.

Another way to visualize this, and also to gain some insight on what the Universe is expanding into, is to consider a hypothetical universe consisting of the two-dimensional surface of a balloon, with stickers (galaxies) attached to it.

Our own Universe might be the three-dimensional analogue of a balloon, where volume wraps around a fourth, unreachable spatial dimension, and there is no unique center within our physically accessible dimensions.

By measuring the distances of many galaxies and calculating their recession speeds from spectra, astronomers have determined the current expansion rate of the Universe fairly accurately: $H_0 = 73 \pm 7$ km/s/Mpc. Thus, a galaxy 10 Mpc away (about 32.6 million light-years) recedes from us at about 730 km/s, according to Hubble's law, $v = H_0 d$. The distances of galaxies were measured using a variety of techniques, and the results are quite consistent.

■ **Above:** *A hypothetical one-dimensional expanding universe. (The balls don't expand, just as gravitationally bound galaxies don't expand.) Fix your attention on one of the balls. As we stretch the rubber tube, all of the balls recede from our chosen ball, and the more distant balls recede faster than the nearby ones because there is more expanding space between distant balls than between nearby ones. But this same view would be seen if you were to fix your attention on any other ball; none of them can claim to be in the unique center. Image courtesy Alex Filippenko.*

■ **Below left:** *A hypothetical two-dimensional expanding universe. (The stickers, which don't expand, represent galaxies; they should really be embedded in the material of the balloon.) You can go forward and backward, or left and right, or any combination of those two motions – but you cannot go inside or outside the balloon. As the balloon expands, each sticker moves away from every other sticker with a speed that is proportional to its distance. Each sticker thinks it is the center of the universe. But actually, the center of the balloon is in a mathematically describable, but physically inaccessible, dimension – it is **not** part of the balloon (the universe) itself! Also, the expansion occurs into this extra dimension. Image courtesy Alex Filippenko.*

The Possible Fate of the Universe

However, that is not the end of the story. We expect the expansion rate to be changing with time, simply because galaxies gravitationally pull on each other, thereby resisting the expansion of space to some degree. Think of an apple tossed into the air: it slows down in its upward journey and then comes back down. Similarly, if there is much mass per unit volume (that is, the density is high), the Universe should slow down rapidly and ultimately halt its expansion, and then reverse it and begin contracting, ultimately collapsing to a 'Big Crunch' or 'gnaB giB' (the opposite of the Big Bang). On the other hand, one could, in principle, heave an apple so fast that (neglecting air resistance) it never falls back to the Earth; its speed would exceed Earth's escape velocity. Applying this concept to the Universe, one concludes that if the density is sufficiently low, the Universe will expand forever, because there is too little mass, too little gravity, to ever halt the expansion. However, even an eternally expanding Universe would gradually slow down its expansion, approaching a constant non-zero expansion speed as time approaches infinity.

Thus, by observing the past history of expansion, one can predict the fate of the Universe: if the Universe has been slowing down rapidly, it will eventually collapse, but if it has been slowing down very little, it will expand forever. To look back into the past, astronomers

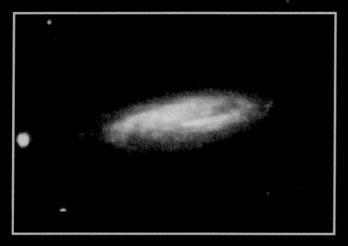

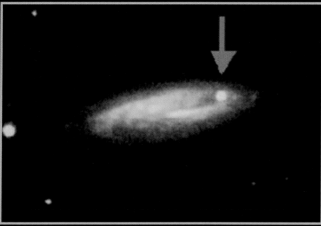

measure galaxies billions of light-years away, because we see them as they were billions of years ago. (Light travels fast, but not infinitely fast; even the nearest stars are typically tens or even hundreds of light-years away, with a light-year being the distance light travels in one year – about 9.5 million million kilometers.)

How do astronomers measure the distances of galaxies? Well, one finds a star whose observed properties nearly exactly match those of a nearby star in our own Milky Way Galaxy. Knowing the distance of the nearby star (through a variety of techniques), and measuring the apparent brightness of that star, we can figure out its true power, or luminosity. Then, by comparing the apparent brightness of the star in the distant galaxy with the known luminosity of the nearby star, we can use the inverse-square law of light to calculate the distance of the distant star, and hence of its home galaxy. This is similar to estimating the distance of an oncoming car at night: you compare the apparent brightness of its headlights with the true brightness (luminosity) of the headlights of a nearby car at known distance.

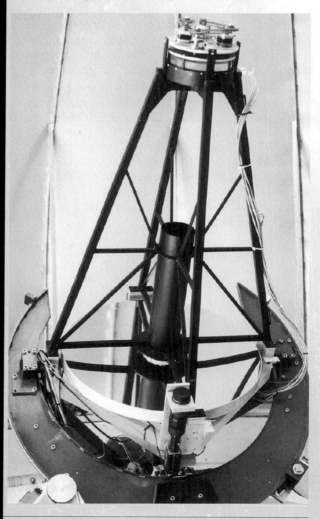

■ *Above: Images of the relatively nearby galaxy NGC 7541 (about 110 million light-years away), before (top) and after (bottom) the supernova (exploding star) SN 1998dh, marked with an arrow, became visible. It shines with the power of about 4 billion Suns. Other stars in the images are normal, foreground stars in our Milky Way Galaxy. The images were taken with a charge-coupled device (CCD) attached to the Katzman Automatic Imaging Telescope (KAIT). Supernovae are named in order of discovery in a given calendar year, from A through Z, continuing to aa through az, then ba through bz, and so on. Image courtesy Weidong Li and Alex Filippenko.*

■ *Right: The Katzman Automatic Imaging Telescope (KAIT), whose primary mirror has a diameter of 0.76 m. The author's Lick Observatory Supernova Search is conducted with KAIT; over the past decade, this has been the world's most successful search for nearby supernovae, generally finding about 80 per year. KAIT automatically takes digital CCD photographs of galaxies (about 7000 per week), compares the new images with the old ones of the same galaxies, and looks for 'new' stars (supernova candidates) in the most recent images. Undergraduate students examine the supernova candidates found by KAIT and reject the false alarms, such as asteroids, or high-energy charged particles (cosmic rays) that hit the detector. In this way, the students participate in useful research quite early in their careers. Image courtesy Weidong Li.*

Supernovae to the Rescue

But this procedure works with the most powerful normal stars only for relatively nearby galaxies, in which we can see individual stars. How can one accurately measure the distances of distant galaxies, whose individual stars are very faint and smeared together? It seems like a hopeless task, until one realizes that sometimes stars can violently explode and become up to 4 billion times as powerful as the Sun!

If one can calibrate the power of such an exploding star, or 'supernova,' in a relatively nearby galaxy of known distance, then one can compare this power with the apparent brightness of a supernova in a distant galaxy, thereby determining its distance.

Of course, one has to make sure that the two supernovae are the same type; there are several varieties of exploding stars, and they don't all have the same peak power. The generally most powerful ones, most of which achieve nearly the same peak power, are known as Type Ia supernovae – they don't have hydrogen lines in their spectra, and they are thought to be produced by white dwarf stars (retired stars made of 'degenerate matter') that steal enough gas from a companion star to reach an unstable mass limit, exploding as a runaway chain of nuclear reactions. Far in the future our Sun will become a white dwarf, but it won't explode because it does not have a companion star from which to steal gas and reach the unstable limit of 1.4 solar masses (known as the 'Chandrasekhar limit').

To calibrate the peak power of Type Ia supernovae, and determine whether there are slight differences among them, one has to observe many such supernovae in relatively nearby galaxies. But supernovae are rare; a typical large galaxy like the Milky Way produces only about two supernovae of *all* types per century. It would be foolish to observe one, and only one, galaxy for decades, searching for a supernova; it's much wiser to search in thousands of galaxies, because your odds of finding supernovae among those galaxies improve. Indeed, my team at the University of California, Berkeley, has a robotic telescope at Lick Observatory that searches for supernovae and measures how they brighten and fade with time.

Through such studies, astronomers have come to a good understanding of the observed properties of nearby Type Ia supernovae. Now let's see how the comparison of distant and nearby supernovae leads to a stunning conclusion about the expansion history of the Universe.

Cosmic Acceleration:
A Surprising Twist

In the early- to mid-1990s, two teams started discovering and measuring Type Ia supernovae at redshifts of about 0.5 – roughly 5 billion light-years away. The first, led by Saul Perlmutter of the Lawrence Berkeley National Laboratory at the University of California, Berkeley, was called the 'Supernova Cosmology Project'; I was part of this team for a few years. The second, led by Brian Schmidt of the Australian National University, was called the "High-z Supernova Search Team," where 'z' stands for redshift. I eventually left Perlmutter's team to join Schmidt's team whose members were largely responsible for calibrating in detail the luminosity of nearby Type Ia supernovae; primarily because of differences in scientific style, but to this day I am glad that there were two teams working on the same topic: the competition accelerated the pace of the work, and led to more careful research since neither team wanted to be viewed as being sloppy.

The teams searched for distant supernovae in the same way: they used large telescopes (such as the 4-meter Blanco telescope at Cerro Tololo Inter-American Observatory, in Chile) equipped with wide-angle cameras to take images down to very faint limits, revealing thousands of individual galaxies (qualitatively similar to what is shown in the Hubble Ultra Deep Field). By re-imaging the same fields in the sky after a few weeks, they could discover distant supernovae

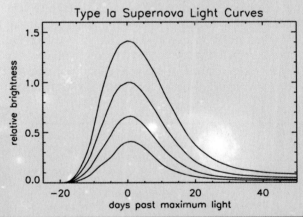

Type Ia Supernova Light Curves

■ **Above:** *Idealized light curves (plots of brightness vs. time) of Type Ia supernovae. Studies of these objects in relatively nearby galaxies show that most of them have nearly the same peak luminosity, the intrinsically more powerful supernovae take longer to brighten and fade (say, from half of their peak brightness back down to half of their peak brightness) than the less powerful ones. By measuring the light curve of a distant supernova and comparing with the known relationship between light-curve shape and peak luminosity, we can calibrate each individual supernova. Image courtesy Saurabh Jha and Alex Filippenko.*

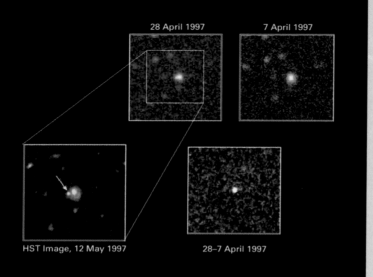

28 April 1997 7 April 1997

HST Image, 12 May 1997 28–7 April 1997

in 'batches,' many at a time (since up to 100,000 galaxies were being monitored, some of which surely produced supernovae during the intervening time interval), though a process of digital subtraction and comparison.

Time for follow-up observations could be scheduled on other large telescopes, including the 10-meter diameter Keck telescopes in Hawaii (which I used to spectroscopically confirm the supernovae) and the 2.4-meter Hubble Space Telescope (to obtain accurate measurements of the supernovae brightening and fading with time).

By late 1997, both of the teams were grappling with a very strange result: the high-redshift supernovae appeared fainter than expected, suggesting that their distances were larger than expected in a decelerating universe, or even in one that was expanding at a constant speed.

Unless something else was affecting the measured apparent brightness of the supernovae, the natural conclusion was that the expansion rate of the Universe is accelerating, speeding up with time instead of slowing down as expected due to normal gravity! This initially seemed like

nonsense, and I had quite a few discussions with my postdoctoral research fellow, Adam Riess, who had been analyzing the data for our High-z Supernova Search Team, but he could not find any error in his analysis after doing many cross checks. Team leader Brian Schmidt also worked tirelessly to figure out what was wrong, but he, too, kept getting measurements very similar to Adam's.

Let me give an analogy to help clarify this result. Suppose that the expansion of the Universe is represented by an apple thrown upward from my hand, and the age of the Universe is just 1 second. The apple reaches a certain distance from my hand in 1 second – but the apple was slowing down during its upward journey from my hand, so the measured distance would have been larger if the Earth's mass (and hence gravity) were weaker (there would have been less deceleration). If the mass of the Earth were zero, so there was no gravitational attraction between the Earth and the apple, the apple would not slow down at all, and its distance after 1 second would be even larger. Well, the supernova brightness measurements, in the context of the apple analogy, implied that the apple is at a greater distance than it could have reached even if it were not slowing down at all. Thus, it must have been speeding up! The supernovae, and the galaxies in which they were located, had reached a greater distance then they could have in a non-decelerating universe, so the Universe is actually accelerating in its expansion, due to some hitherto unknown cosmic 'antigravity' effect. (We use the term 'antigravity' hesitantly, because the 'stuff' causing the accelerated expansion almost certainly cannot be attached to cars, for example, to make them levitate over traffic jams! Instead, the effect is probably associated with space itself, as I will describe below.)

■ **Above:** *Small portion of a relatively wide-angle CCD image (similar to the Hubble Ultra Deep Field), obtained on 7 April 1997 (top right) and again on 28 April 1997 (top left). Upon subtraction of the 7 April image from the 28 April image, a supernova candidate is visible (bottom right), along with some noise (any measurement process has noise associated with it). A spectrum showed that the object is indeed a supernova (SN 1997cj), with redshift z = 0.50. A Hubble Space Telescope image obtained on 12 May 1997 (bottom left) shows it clearly. Image courtesy Brian P. Schmidt and the High-z Supernova Search Team.*

■ **Right:** *Wide-angle (top) and close-up (bottom, marked with arrows) Hubble Space Telescope images of three distant Type Ia supernovae (left: SN 1997cj, z = 0.50; center: SN 1997ce, z = 0.44; right: SN 1997ck, z = 0.97). At peak brightness, they appear very faint – fainter than expected (given their redshifts) in a universe whose expansion rate slows down with time. They are even too faint to be consistent with a universe whose expansion rate is constant. Image courtesy Peter Garnavich and the High-z Supernova Search Team, and NASA.*

Now, you might object that the conclusion regarding distances was based on an assumed age of the Universe. If the Universe were 2 seconds old, rather than just 1 second as assumed, then the apple (i.e., the galaxies hosting the supernovae) would have reached a distance greater than expected without any acceleration. If this is what you were thinking, good try – but it turns out not to be a valid objection. When the data are analyzed and rigorously interpreted in detail, the conclusion that the Universe is accelerating ends up being *independent* of the assumed age of the Universe. Technically, what we find is that each supernova is at a greater distance than what had been expected based on its redshift, in a non-accelerating universe. Thus, the expansion of the Universe must be accelerating.

One must also account for possible non-cosmological effects that could affect the apparent brightness of high-redshift supernovae, and the High-*z* Team spent much time pondering these and determining their likely magnitude. For example, if substantial interstellar gas and especially dust were present along the line of sight to the supernovae, they would look dimmer than expected. But gas and dust tend to extinguish blue light more than red light; witness the color of the setting Sun, which is more yellow, orange, or even red than the Sun seen high in the sky, because the shorter-wavelength light was selectively filtered out by Earth's atmosphere. By measuring the colors of high-redshift Type Ia supernovae (first taking into account the redshift itself), we could tell that the supernovae were not dimmer due to interstellar dust. Another possible effect is that supernovae long ago were intrinsically less powerful than recent, local Type Ia supernovae. However, by carefully examining the spectral properties of the high-redshift supernovae, we ruled out this alternative fairly convincingly.

Thus, although we were not absolutely certain of our conclusion, by February 1998 we were reasonably convinced that either the Universe is accelerating in its expansion, or some very subtle effect is fooling us – but presumably we would learn something interesting about the Universe even in the latter case, so it was reasonable to announce our results and their associated statistical (and systematic) uncertainties.

I was privileged to present the High-z Supernova Search Team's results for the first time at a cosmology meeting near Los Angeles, California, 20-22 February 1998. I stated that the High-z Team had convincing evidence that the expansion of the Universe is accelerating, based on our analysis of the high-redshift and low-redshift Type Ia supernovae. An excited, but cautious, buzz could be heard among the assembled astronomers and physicists in the audience; this was a very exciting, unexpected discovery!

Saul Perlmutter's talk preceded mine, and (much to my relief) he did not present such solid evidence; the Supernova Cosmology Project's data indicated that the density of the Universe was sufficiently low that the Universe would expand forever, but the uncertainties were still too large to ascertain whether the expansion was accelerating or decelerating. His team's data were certainly consistent with acceleration, but given the uncertainties at the time, they were also entirely consistent with the more conventional, expected result of no acceleration. Hence, he did not emphasize the acceleration as much as I did, and the resulting press coverage (initially through James Glanz's report on page 1298 in the 27 February 1998 issue of *Science* magazine) focused primarily on the High-*z* Team's results.

Moreover, the High-*z* Team's scientific paper was submitted for publication in March 1998 and published in September 1998, whereas the Supernova Cosmology Project's paper was submitted in September 1998 and published in June 1999. Nevertheless, the fact that two independent teams (I was not associated with the data analysis done by the Supernova Cosmology Project) had come to essentially the same conclusion nearly simultaneously made understandably incredulous astronomers take more careful note of the results than they would have had there been just one team. The possibility of an accelerating expansion rate became the focus of attention at a May 1998 meeting of physicists and astronomers in Chicago; there, they carefully examined the available evidence and began to consider in detail potential causes for the effect.

Einstein's Biggest Blunder?

By the end of 1998, the editors of *Science* magazine proclaimed the two teams' discovery of the accelerating expansion of space to be the single most important scientific breakthrough of 1998. The cover of their 18 December issue shows the caricature of Einstein symbolically blowing the Universe out of his pipe, and that Universe subsequently expands faster with time, rather than more slowly. But Einstein looks doubly surprised, because in his hand he holds a sheaf of papers with the equation $\Lambda = 8\pi G$ times the density of the vacuum (G is Newton's universal constant of gravitation, and Λ is the Greek upper-case 'lambda'). Now, you might think that the vacuum, being sheer emptiness, doesn't have a nonzero density, yet in

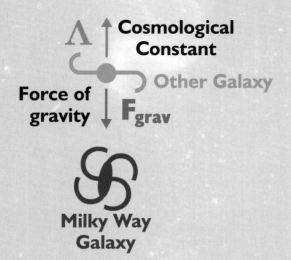

Λ ↑ Cosmological Constant

Other Galaxy

Force of gravity ↓ F_grav

Milky Way Galaxy

1917 Einstein had suggested the possibility of a gravitationally repulsive, positive energy density – exactly the sort of thing that is needed to explain the accelerating expansion of the Universe!

Einstein had reasoned that the Universe is static, an aesthetically pleasing (to himself) possibility that seemed supported by observations at that time. But to have a static universe, when in fact the gravitational attraction of all objects for each other should cause contraction, there had to be a long-range repulsive effect acting in the direction opposite gravity, and with exactly equal magnitude or size.

Thus, in 1917, shortly after developing his general theory of relativity, Einstein introduced Λ, the 'cosmological constant,' corresponding to a nonzero repulsive energy that exactly cancels attractive gravity on the largest scales. This 'fudge factor' was not mathematically incorrect; in fact, it made the equations more general, to some degree. But its magnitude had to be artificially tuned to produce a perfectly static universe – an unlikely condition that made the equations lose their aesthetic appeal. Moreover, there was no other experimental, laboratory evidence for repulsive energy, and it had no known physical origin. Einstein was not fond of the cosmological constant, but he felt it was needed in order to explain the apparently static nature of the Universe.

After Edwin Hubble's 1929 discovery that the Universe isn't static after all, the entire physical and philosophical motivation for the

Above: *Schematic illustrating Einstein's 'cosmological constant' (Λ) and a static universe. The attractive force of gravity between the Milky Way Galaxy and another galaxy (downward pointing arrow) is exactly balanced by a repulsive effect Λ (upward pointing arrow), of unknown physical origin. But if the up arrow dominates over the down arrow on the largest scales, then the expansion of the Universe actually accelerates with time; the Universe is not static in this case. Image courtesy Alex Filippenko and Frank Serduke.*

cosmological constant vanished. After all, the Universe could have begun its existence in an expanding state (the Big Bang), and no strange forces are needed to continue that expansion. Einstein renounced the cosmological constant, saying that there was no longer any motivation for retaining it. He was sad that he had not predicted the dynamic nature of the Universe, as some other theoretical physicists had done. Anecdotally, according to physicist George Gamow, Einstein called the cosmological constant the 'biggest blunder' of his career.

What have we done, about seven decades later? We've reincarnated the concept of a repulsive energy in the vacuum, not to make a static universe, but rather one whose expansion accelerates over the largest size scales. On Earth, and in our Solar System, and in our Milky Way Galaxy, and in our Local Group of galaxies, attractive gravity dominates – but over distances of hundreds of millions of light-years, the cosmological constant (or something like it) dominates, causing space to expand faster and faster with time. Thus, Einstein's 'biggest blunder' might actually have been one of his greatest triumphs; the concept of repulsive energy is not incorrect, only the precise magnitude that Einstein ascribed to it was wrong.

Early Cosmic Deceleration

The discovery of the accelerating expansion of the Universe in 1998 was based on observations of only about 50 Type Ia supernovae. (Since then, several hundred additional objects have been observed, amply confirming and refining the initial results.) These supernovae were at a typical redshift of 0.5 – that is, about 5 billion light-years away. This means that the Universe has been accelerating for at least the past 5 billion years. But it was of interest to see what the expansion was doing in the first 9 billion years of the Universe's roughly 14 billion year existence. To find out, we needed to discover and measure Type Ia supernovae at ever greater distances. A team led by Adam Riess, in which several of the past members of the High-z Supernova Search Team (including myself) participated, found such supernovae with the Hubble Space Telescope and measured their apparent peak brightness.

Analysis of the data revealed that the Universe was, in fact, decelerating during its first 9 billion years, and started accelerating only about 5 billion years ago! The transition from deceleration to acceleration is a change in the deceleration, known mathematically as a 'jerk'; thus, we found that the Universe went through a jerk about 5 billion years ago. (Amusingly, the

■ *A cluster of galaxies, Abell 2218, imaged with the Hubble Space Telescope. Clusters such as this one require large amounts of invisible 'dark matter' to keep them gravitationally bound; otherwise, they would rapidly fly apart and destroy themselves. Image courtesy NASA, Andrew Fruchter, and the Early Release Observations Team (STScI).*

resulting *New York Times* headline read, "A Cosmic Jerk That Reversed the Universe" – and right below it was a photo of Adam Riess! A few people jokingly asked me, "Hey, who is this jerk you work with?")

Why might we expect the Universe to have gone through such a transition? Well, if the repulsive energy is a constant property of space itself, independent of the size of the Universe, then the more space there is between galaxies, the greater is the cumulative repulsion. Long ago, when galaxies were close together, their gravitational attraction for each other was stronger than it is now, and the repulsion was weaker. But as the Universe expanded, the distances between galaxies increased, and their gravitational attraction weakened while the repulsion grew stronger (being a cumulative effect over large distances). Eventually, the typical distances between galaxies became so great that repulsion began to dominate over attraction, and deceleration turned to acceleration.

Dark Energy: The Dominant Stuff of the Universe

What could be the 'stuff' that currently causes space to expand more rapidly with time? Certainly it is not ordinary, luminous matter of which stars and galaxies consist; ordinary matter has a gravitationally attractive effect. It is also not the 'dark matter' that must dominate the mass in galaxies and clusters of galaxies.

For example, if clusters of galaxies were *not* filled with gravitationally attractive (but invisible) matter, they would rapidly be destroyed, because the galaxies would have no reason to remain so close together (an argument first proposed in the 1930s by the brilliant, but arrogant and abrasive, astrophysicist Fritz Zwicky of Caltech). While we don't know what about 90 percent of the dark matter consists of (probably some sort of fundamental particles left over from the Big Bang, generally known as 'weakly interacting massive particles' or WIMPs), we *do* know that it is gravitationally attractive, and hence cannot cause the observed acceleration of the Universe.

The agent causing the acceleration must be some sort of new, exotic stuff – and for want of a better term, it is now known as 'dark energy.' This nomenclature is, to some extent, regrettable.

Before Supernova   Before Supernova   Before Supernova

■ *Left: Hubble Space Telescope images of three of the most distant Type Ia supernovae ever found, along with their host galaxies prior to the explosions. These objects are seen as they were when the Universe was roughly half its present age. They are brighter than expected if the Universe had always been accelerating; therefore, the expansion must have been decelerating when the Universe was young. Image courtesy NASA and Adam Riess.*

After all, the most famous equation in science is Einstein's $E = mc^2$, and this equation perhaps implies that 'dark energy' *(E)* is equivalent to 'dark matter' *(m)*. Nothing could be further from the truth; dark energy pushes, whereas dark matter pulls... and while dark energy has a normal, positive energy density, it pushes because it appears to have 'negative pressure,' which in the general theory of relativity leads to an overall repulsive effect.

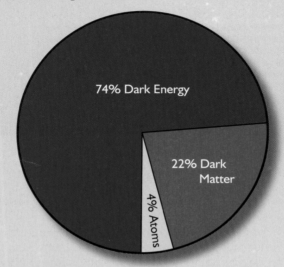

Dark energy is the dominant component of the mass-energy density of the Universe, constituting about 74 percent of the overall total. Of the remainder, 22 percent is dark matter of unknown type (possibly WIMPs), and 4 percent normal matter made of protons and neutrons. Thus, although dark energy is negligible in our Solar System, Galaxy, and Local Group of galaxies, it does dominate over distances of billions of light-years.

While the original direct evidence for an accelerating Universe was reported in 1998 from studies of Type Ia supernovae, and one could worry about whether the adopted technique might be flawed, there are now several largely independent lines of evidence that suggest the presence of repulsive dark energy. Detailed observations of the cosmic microwave background radiation, the afterglow of the Big Bang, provide some of the strongest independent evidence. In particular, data from the Wilkinson Microwave Anisotropy Probe (WMAP) convincingly show that the inferred overall mass-energy density of the Universe requires the presence of a new, dominant component of energy. While each of the independent methods might be flawed, it seems quite unlikely that all of them would conspire to give the same wrong answer; instead, they seem to all be consistent with a repulsive dark energy constituting roughly three-quarters of the total mass-energy density of the Universe. It is fitting that Saul Perlmutter, Adam Riess, and Brian Schmidt were awarded the prestigious 2006 Shaw Prize (sometimes called the "Nobel Prize of the East") in Astronomy for their roles in the discovery of cosmic acceleration and dark energy. See **http://www.shawprize.org/en/laureates/2006/astronomy/index.html.**

In honor of dark energy, my wife Noelle thought up an appropriate personalized license plate for my car.

The physical nature of dark energy is often agreed to be the most important observationally motivated, unsolved problem in all of physics. We really have no compelling model for it at the present time. However, it is believed that an understanding of dark energy will require the long-sought unification of the two great pillars of modern physics: general relativity, which deals with large masses and large distances (neutron stars, galaxies, clusters of galaxies), and quantum mechanics, which deals with very small masses and small distances (subatomic particles, atoms, molecules). Each of these theories works extremely well in its domain of applicability, but they are completely incompatible (indeed, at war with each other) when one tries to use both theories to study a large amount of mass in a small volume (such as the singularity inside a black hole). Any purported unified theory (such as a specific 'string theory') must be compatible with the presence of dark energy, or it can be eliminated from consideration. Thus, dark energy offers an enormously valuable clue to the structure of a truly unified theory.

■ *Above: A pie chart showing the composition of the Universe, as revealed through studies of Type Ia supernovae, the cosmic microwave background radiation, and the large-scale structure of the Universe. Only 4 percent of the Universe consists of normal atoms, most of which are hard to detect, and hence are also relatively 'dark'; only about 0.4 percent of the Universe consists of matter that is easily visible in the form of stars and glowing gas. We don't understand the composition of about 96 percent of the contents of the Universe! Image courtesy NASA/WMAP Science Team.*

■ *Above right: The license plate of the author's car, in honor of dark energy and the accelerating universe. Can you figure out how the license plate reads "dark energy"? Image courtesy Noelle Filippenko and Alex Filippenko.*

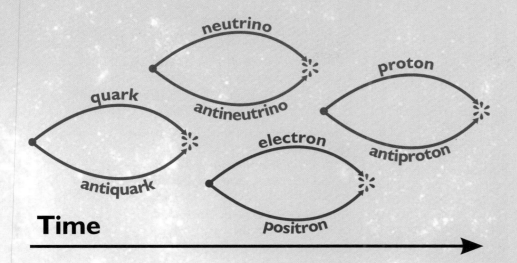

Time

One of the simplest ideas is that dark energy results from 'quantum fluctuations' in the vacuum – tiny bits of energy that pop into and out of existence over minuscule time scales in the form of particle-antiparticle pairs such as an electron and positron, or anti-electron.

Such fluctuations are a natural consequence of the Heisenberg Uncertainty Principle, one of the foundations of quantum mechanics. Though they cannot be directly observed, their effects can be measured; for example, the energy levels of electrons in atoms are ever so slightly affected by quantum fluctuations. A major problem with quantum fluctuations being the source of dark energy, however, is that one expects a far, far greater *density* for the dark energy, compared with what astronomers actually measured. So, theorists had always assumed that there are negative-energy fluctuations that exactly balance the positive-energy fluctuations, leaving no net energy. But if the fluctuations don't quite cancel out, leaving a bit of positive energy, it would have the desired effect of accelerating the expansion of space.

At the present time, the observations of distant Type Ia supernovae are most consistent with the quantum fluctuation ('vacuum energy') hypothesis, although theoretical physicists don't like it because of the required, yet unlikely, nearly perfect (but slightly imperfect) cancellation. There is also no explanation for why the overall contribution of dark energy is *roughly* comparable to that of dark and normal matter at the present time; the ratio 74%/26% ≈ 2.8 is nearly unity, yet it could have been essentially any number far from 1.0, such as 10^{50} or 10^{-15}.

Alternative hypotheses abound, but seem somewhat contrived, and do not currently yield better agreement with the observations. One generic class of models is called "quintessence." The basic idea is that the vacuum actually has a new kind of field associated with it (perhaps because two forces remained 'unified' together at a time when they should have been separate), and the strength of this negative-pressure, repulsive field can change with time (that is, it isn't a constant density like the quantum fluctuations). A qualitatively similar, but quantitatively much stronger, field of this kind is postulated to have launched the initial, extremely rapid, exponential growth of the Universe when its age was a tiny fraction of a second – the era of 'inflation' when the Universe went 'bang' from a tiny entity to something much larger than the observed radius of the Universe. (Space expanded much faster than the speed of light during this time, but such expansion is not a violation of Einstein's special theory of relativity because no signal carrying information can be transmitted by superluminally expanding space.) Although we are currently beginning to enter a new inflationary epoch, it might someday end (when the field changes form), as did the initial era of inflation of the early Universe.

The way to discriminate among the competing models is to observe, in much greater detail, the expansion history of the Universe. Different hypotheses predict different expansion histories, though in many cases the differences are subtle. Several groups of astronomers are now finding hundreds of Type Ia supernovae, spanning a wide range of redshifts, and measuring their light curves, with the hope of quantifying more accurately the expansion rate as a function of time. Spacecraft dedicated to studies of thousands of supernovae are being planned, though it is not yet clear that they will be funded because of the tight NASA budget.

Conclusion

So, what will be the fate of our Universe? We don't really know, because the physical nature (and future behavior) of dark energy is not yet understood. If the dark energy remains repulsive, then the Universe will expand faster and faster with time forever – a 'runaway universe.' Should you want to see clusters of galaxies with your very own eyes, go look through a telescope relatively soon, within the next few tens of billions of years, before those clusters are whisked away to such great distances that they become too faint! The runaway universe will rapidly become cold, dark, and dilute – an ending 'in ice,' so to speak. On the other hand, if the dark energy someday changes sign and becomes gravitationally attractive, the Universe could eventually collapse into a Big Crunch ('gnaB giB'), becoming hot, bright, and compressed – an ending 'in fire.' However, based on what we know now, the Big Crunch won't occur for at least another 40 billion years, and perhaps much longer.

Although the American poet Robert Frost could not have known about repulsive dark energy, he apparently did know the two possible fates for the Universe: ultimate collapse (an ending in fire), and eternal expansion (an ending in ice). Recall his well-known poem (1920), *Fire and Ice:*

Some say the world will end in fire,
Some say in ice.
From what I've tasted of desire
I hold with those who favor fire.
But if I had to perish twice,
I think I know enough of hate
To know that for destruction ice
Is also great
And would suffice.

We see that Robert Frost would prefer the fiery collapse of the Universe. But if he had to die twice, eternal expansion and an ending in ice would be acceptable – and that's kind of appropriate, given his name, Robert *Frost.*

In summary, the discovery of dark energy and the currently accelerating ('runaway') universe is one of the most important scientific breakthroughs of the past few decades. It is revolutionizing physics, showing that gravity has another, repulsive, aspect to it, when considered over the largest scales in the Universe. The presence and properties of dark energy are a crucial test for any 'theory of everything' that attempts to unify quantum mechanics and general relativity, providing the fundamental framework of physics. Stay tuned for further observational and theoretical developments in this field!

In July 2007, the members of the two international teams that discovered the accelerating expansion of the universe – the Supernova Cosmology Project and the High-z Supernova Search Team, were awarded the 2007 Gruber Cosmology Prize. The author, Alex Filippenko, is one of the authors in both of the original research papers.

For further information, consult these websites:
High-z Supernova Search Team:
http://cfa-www.harvard.edu/supernova//HighZ.html
Supernova Cosmology Project:
http://panisse.lbl.gov/
Katzman Automatic Imaging Telescope:
http://astro.berkeley.edu/~bait/kait.html
Professor Ned Wright's cosmology tutorial:
http://www.astro.ucla.edu/~wright/cosmolog.htm
Wilkinson Microwave Anisotropy Probe:
http://map.gsfc.nasa.gov/
Shaw Prize lectures:
http://www-int.stsci.edu/~ariess/shawlecture.pdf
http://www.keckobservatory.org/support/magazine/2007/mar/media/schmidt_shaw_lecture.pdf

Image courtesy Steve McConnell

Alex Filippenko received his Ph.D. in Astronomy from Caltech in 1984 and joined the UC Berkeley faculty in 1986. He has coauthored over 500 scientific publications and has won numerous prizes for his research. He has won the top teaching awards at Berkeley, and in 2006 was named the Carnegie/CASE National Professor of the Year among doctoral institutions. He has produced a 96-lecture astronomy video course with The Teaching Company and has coauthored an award-winning astronomy textbook.

THE END

Printing: Mercedes-Druck, Berlin
Binding: Stein+Lehmann, Berlin